The
Dangers
Of
5G

Claudia Drake

ISBN 13: 9798610041612

Disclaimer

The information provided by The Dangers of 5G ("we," "us" or "our") is for general information only. All information is provided in good faith, however we make no representation or warranty of any kind, express or implied, regarding the accuracy, adequacy validity, reliability, availability or completeness of any information. Under no circumstance shall we have any liability to you for any loss or damage of any kind incurred as a result of the use of any information provided. Your use of the book and your reliance on any information is solely at your own risk.

The book may contain links to other sites or content belonging to or originating from third parties or links to websites and features in banners or other advertising. Such external links are not investigated, monitored, or checked for accuracy, adequacy, validity, reliability, availability or completeness by us. We do not warrant, endorse, guarantee, or assume responsibility for the accuracy or reliability of any information offered by third-party websites linked through the book or any website or feature linked in any banner or other advertising. We will not be a party to or in any way be responsible for monitoring any transaction between you and third-party providers of products or services.

This book cannot and does not contain medical/health advice. The medical/health information is provided for general information and educational purposes only and is not a substitute for professional advice. Accordingly, before taking any actions based upon such information, we encourage you to consult with the appropriate professionals.

Table of Contents

Forward

5G is a new concept in cellular technology. It is changing the world we live in. We are at the beginning of the transformation, and how we allow it to move forward will affect generations to come.

5G is so new, it is being invented as it is being implemented. It will literally change the landscape around us, with "small cell" antenna being mounted every 1000 feet or so--right in front of your house.

The Federal Communications Commission, the FCC, is allowing the cellular industry to run the show. The Telecommunications Act of 1996 was put in place to promote competition and reduce regulation in order to secure lower prices and higher quality services for telecommunications consumers, and to promote rapid deployment of new telecommunication technologies.

But in creating the rules of deployment, the FCC has ignored local government and individual input. They have removed local government's right to control where and when the new 5G technology is installed.

The FCC has also ignored the monumental amount of research by independent scientists on the dangers of the past and current levels of radio frequency electromagnetic fields/frequencies (EMFs). The parameters of cellular technology have changed tremendously since 1996 with the advent of 5G, yet the safety guidelines have not taken this into consideration.

While the World Health Organization (WHO) has found that cellphones might possibly cause some brain cancers, there are many research studies that have found a more definite link. In a ridiculous lack of logic, some accepted studies deny cell phones are a threat, while, at the same time, admitting there is an association of brain cancer in the case of "heavy use."

5G radiation has been shown to cause changes in the body, blood and bone marrow composition, the eyes, heart, immune system, inflammatory responses, circulatory system, nervous system, and antibiotic response.

5G, using millimeter wavelengths (MMWs) does not replace electromagnetic frequencies (EMFs). They will be added to it. We will receive both MMW and EMF emissions at the same time.

We, as a society, no longer use our cell phones just to talk on the phone when we are away from home. Our use has escalated exponentially. Many people have eliminated their landline phone. The cell phone is used as a mobile computer with access to the Internet, social media, news, games, and videos.

Children are given their own cell phone at the approximate age of 10 or sooner. Very young children are kept preoccupied with games and songs on their parents' cell phones. And they are even more vulnerable than adults.

When 5G small cell antennas are installed every 1000 feet outside our homes, we will be drenched in millimeter waves, (MMW), and electromagnetic fields, (EMF), 24 hours a day, 7 days a week.

This book presents the studies that show a connection between MMW, EMF and negative health effects. It demonstrates the toxic relationship between the telecommunications industry and the federal agency, the FCC, tasked with the responsibility of keeping the American public safe. It gives you the tools you need to

make an informed decision on the safety of 5 G for
yourself.

5G

5

5G is the "fifth generation" of cellular technology. It promises lightning-fast speeds and the ability to power new technologies like self-driving cars, remote surgery, and virtual reality experiences. It is not just an extension of the present 4G, with faster speeds; it is an entirely new concept in the deployment of data delivery. Some believe 5G could be one of the most important developments in human history. But an undercurrent to all the things we can do with 5G is concern about what 5G can do to us.

While the cellular companies are going full steam ahead, not everyone believes we should be going so fast. There are concerns the very high frequency spectrum known as millimeter wavelengths (MMW) used in deployment to make 5G a reality could pose adverse health effects for the public.

How is 5G different from the prior technology?

Mobile devices will be able to send and receive information in less than one-thousandth of a second, appearing instantaneous to the user. But to accomplish these speeds, the rollout of 5G requires new technology and infrastructure.

What frequencies are used for what? The low frequencies (3 HZ – 300 KHz) are used for electrical power line transmission (60 Hz in the US) as well as maritime and submarine navigation and communications.

Medium frequencies (300 KHz – 900 MHz) are used for AM/FM/TV broadcast in North America.

Lower microwave frequencies (900 MHz – 5 GHz) are used for telecommunications such as microwave devices/communications, radio astronomy, mobile/cell phones, and wireless LANs.

Higher microwave frequencies (5 GHz – 300 GHz) are used for radar and proposed for microwave Wi-Fi.

Terahertz frequencies (300 GHz – 3000 GHz) are used increasingly for imaging to supplement X-rays in some medical and security scanning applications (Kostoff and Lau 2017 in Kostoff 2019)).

The new technology is millimeter waves (MMW), using frequencies from 30 up to 300 GHz (gigahertz). These are 10 to 100 times higher than the radio waves used today for 4G and Wi-Fi networks. MMW frequencies are typically in the 24-90 GHz range.

Current radio waves are measured in centimeters and inches, however, 5G electromagnetic radio waves are called millimeter waves because their wavelengths vary between 1 and 10 millimeters.

For cell phones, the frequency range is 900-1800 MHz(megahertz—one million cycles per second) for 3G. For 4G, the frequency is 2-8 GHz (gigahertz—one billion cycles per second), and up to 90 GHz carrier frequency for 5G, so far.

Apart from the direct radiation from individual devices such as mobile phones, humans are also exposed to the electromagnetic field (EMF) from basic radio, electric and telecommunication stations. It encompasses radiation from Wi-Fi home and office routers, including situations when Wi-Fi zones from various sources are being overlapped at one particular spot. Each person is potentially exposed to various sources of EMFs at the same time.

The biological effects of multi-source and multi-frequency EMFs have yet to be explored and fully understood. Presently, there is not any available information on the cumulative effect of such combined exposure.

At this point in time, 5G will be an adjunct to 4G; it will not totally replace it. There will be a mix of frequencies and therefore wavelengths. On April 16, 2019, PC Magazine noted, "The actual 5G radio system, known as 5G-NR, isn't compatible with 4G. But all 5G devices in the US, to start, will need 4G because they'll lean on it to make initial connections," before trading up to 5G where it's available.

4G microwave radiation, at 2.4 GHz and 5 GHz, passes through bodies and the energy is absorbed by anything that contains water. 5G penetrates only the outer layers of the skin in humans. The mix of frequencies in cell towers and cell phones will have a mix of skin and body penetration.

Types of radiation can be organized by their levels of power on the electromagnetic spectrum. Bigger wavelengths with lower frequency are less powerful, while smaller wavelengths at high frequencies are more powerful. This spectrum is divided into two distinct categories: ionizing and non-ionizing.

Ionizing radiation, which includes ultraviolet rays, X-rays and gama rays are the harmful forms. Non-ionizing radiation has lower frequencies and bigger wavelengths. Included are radio frequency, or RF, radiation such as FM radio, TV signals and cellphones that use traditional 3G and 4G service.

Microwave and millimeter wavelength radiation, which is one of the key blocks of the spectrum that 5G service will use, is also considered non-ionizing and is purported to not produce the kind of energy that directly damages cells, according to the telecommunications industry. Common devices, such as Wi-Fi routers, garage door openers, airport security scanners and walkie-talkies, use lower frequency microwaves.

But some experts suspect that the radiation from these devices could damage cells via another biological mechanism, such as oxidative stress in cells, which leads to inflammation and has been found to cause cancer, diabetes and cardiovascular, neurological and pulmonary diseases. Oxidative stress occurs if the equilibrium between formation of reactive oxygen species (ROS) and the capacity of the antioxidant system to neutralize them is disturbed.

While the cellular industry claims there is no danger with 5G due to the fact that the millimeter waves are only superficially absorbed by the skin of humans, it has been shown that 4G and low band 5G emissions can cause oxidation of tissues (Yakymenko 2016).

5G millimeter wavelengths reportedly can have an effect through heat (tissue destruction) through a resonant effect of increased vibration in an object the size of the wavelengths, and at low power levels through signaling of skin structures that can affect metabolism, the nervous system, the endocrine system, and the reproductive system. Millimeter wavelengths at high intensity have also been used in military applications in Active Denial Systems (9 MMW for non-lethal crowd control weapons) that create heat to repel persons. (Declassified Military Studies; Active Denial System).

4G can travel dozens of miles in a line of sight if poles are placed high enough. In experiments, 5G can travel a few miles, but it is easily blocked/absorbed by objects, trees, and plants. Therefore, the installation of 5G "small cells" (antennas) are planned for every 300 meters (about 1000 feet) in cities. Typical installations could involve up to 20 to 25 small cells per square mile. They will be placed on utility poles, streetlights, and buildings, and local government will have no say in their deployment.

The short millimeter waves for 5G technology bring along new challenges for public safety. To improve the performance of 5G and "internet of things" (IoT) devices, pulsed phased antenna arrays will be used in cell towers and wireless devices.

This beam forming technology uses a cluster of pulsed microwave antenna beams whose shape and direction can be controlled to have an individual beam pointed at a device to improve its signal, similar to missile control. 5G technology will have dozens of smaller antennas packed in a single cell antenna array. The heat generated by these antennas is a huge concern (Nasim and Kim 2017).

The exact frequencies of MMW desired for the next-generation of high-speed wireless technologies are not yet configured, but industry letters to the Federal Communications Commission (FCC) seek to open all the frequencies up to 100 GHz, with some suggesting even higher frequencies (FCC Letter 5G Americas).

These MMW frequencies will be mixed with current longer microwave frequencies (4G) to achieve integration of systems. At higher power densities, cell tower studies show that symptoms of electrosensitivity occur within about 300 meters of a cell tower (Santini 2002; Zothansiama 2017)—the estimated distance between the installation of smart cell towers/poles for 5G service.

Representative Thomas Suozzi, a Democrat from New York, expressed concerns. "Small cell towers are being installed in residential neighborhoods in close proximity to houses through my district," he said in a letter to the FCC in 2019. "I have heard instances of these antenna being installed on light poles directly outside the window of a young child's bedroom. Rightly so, my constituents are worried that should this technology be proven hazardous in the future, the health of their families and value of their properties would be at serious risk" (Suozzi May 29, 2019 letter to the FCC).

Heat generated is a concern in handheld devices for 5G and is considered the only valid measure of harm by the FCC. No biological cellular alternations are considered (Wu 2015b). This is a problem.

Many studies demonstrate effects well below the heat threshold of current safety standards (Wyde 2016; Sage and Carpenter 2012; EPA 1992; Esmekaya 2011; Grigoriev 2010; Belyacv 2005).

New research by Neufeld and Kuster (2018) highlight the significant tissue heating generated by 5G technology with rapid short bursts of data transfer on a device. This prompted the authors to call for reevaluation of thermal safety standards in addition to biological standards. The researchers state, "The results also show that the peak-to-average ratio of 10,000 tolerated by the International Council on Non-ionizing Radiation Protection (ICNRP) guidelines may lead to permanent tissue damage after even short exposures, highlighting the importance of revisiting existing exposure guidelines."

A recent National Toxicology Program's (NTP) "Carcinogenicity Studies of Cell Phone Radiofrequency Radiation" report finds a significant increase in heart and brain tumors with radio frequency electromagnetic radiation (RF-EMR) exposure (Wyde 2016). This is in addition to the abundance of basic scientific studies that show a health risk associated with exposure to radiofrequencies, especially with long term exposure (Hardell 2013a, 2013b; Adams 2014; Bortkiewicz 2017; Carlberg and Hardell 2017; Hassanshahi 2017; Liu 2014, Levitt and Lai 2010).

Radiofrequencies are absorbed by and pass through living systems that contain water. Pregnant women and children are more vulnerable to developmental harm from microwave radiation due to immature organ systems (Birks 2017; Othman 2017a, 2017b

Engineers vs doctors and scientists

There is a basic conflict on findings for the safety of 5G between 1) engineers, and, 2) doctors and biological scientists. The Institute of Electrical and Electronics Engineers (IEEE) is an industry association primarily composed of engineers, computer scientists, software developers, information technology professionals, physicists, scientists, and allied professionals. It is the world's largest technical professional association. Their input has been pivotal in setting standards for radio frequency exposures.

The IEEE performs its own research and is involved in computer and wireless technology product development. Even in their own literature, they reiterate the fact there is very little research available. In their 2015 report of 5G, "Safe for Generations to Come" (Wu 2015b), the authors state, "At this time, more reports of beneficial effects than detrimental effects from low-level exposure to mmWave radiation appear to exist in the literature, but this area needs to be better understood, and the specific effects need to be demonstrated reproducibly by independent investigators before any potential non-thermal affects are to be considered in determining the regulatory limits on this regime of non-ionizing radiation."

Current Federal Communications Commission (FCC) Guidelines for non-ionizing radiation exposure were developed over two decades ago and are based on heating of tissues over short exposure periods (6 min. for occupational/controlled and 30 min. for public/uncontrolled exposure) (FCC 1997b, 2015, FCC

2013). The FCC's safety regulations don't account for the higher frequencies that 5G services use. 5G is basically untested for safety.

There are no long-term exposure guidelines, nor are there guidelines for low level, non-thermal or biological effects considered in the International Commission on Non-ionizing Radiation Protection (ICNIRP) standards, which are the basis for standards used worldwide (ICNIRP 2009; Hardell 2017).

They appear to be basing their assumption of safety on a majority, rather than a collaborative view of the findings. They are disregarding the negative studies because they say there are fewer of them. One could conclude that so far, there is not enough science to support the safety of this technology. According to the IEEE, and therefore the FCC, heat is still the only standard of harm considered in 5G applications, not biological effects.

In a 2019 review study, "5G Wireless Communication and Health Effects," by Simko and Mattson (2019), 94 *in vitro* and *in vivo* millimeter wave studies were examined. There were no epidemiological studies found to review. It is noted that this study was funded by a grant from Deutsche Telecom, the largest telecom provider in Europe and one of the largest in the world.

The authors state, "The available studies do not provide adequate and sufficient information for a meaningful safety assessment, or for the question about non-thermal effects. There is a need for research regarding local heat developments on small surfaces, e.g., skin or the eye, and on any environmental impact."

They conclude, "In summary, the majority of studies with MMW exposures show biological responses . . . . Regarding the quality of the presented studies, too few studies fulfill the minimal quality criteria to allow any further conclusions" (Simko and Mattson 2019).

The Telecommunications Act of 1996

The Federal Communications Commission (FCC) is an independent government agency responsible for regulating the radio, television, and phone industries. The FCC regulates all interstate communications, such as wire, satellite and cable, and international communications originating or terminating in the US. Congress reportedly passed the Telecommunications Act of 1996 as a means of restructuring the US telecommunications' markets and included language to facilitate rapid deployment of 5G small cell antennas without local government interference..

The FCC's standards set in 1996 reflected the typical amount of use during that time for a 200-pound man. But phones back then were just that — phones. Now with unlimited games, applications, and social media, the average time spent on smartphones is 3 hours and 10 minutes a day. And that's from people of all ages, sizes and genders.

The US Telecommunications Act of 1996, and the Spectrum Act of 2012, limit local government's ability to control infrastructure for cellular communications and simultaneously facilitates industry deployment. The Act states, "No State or local government or instrumentality thereof may regulate the placement, construction, and modification of personal wireless service facilities on the basis of the environmental effects of radio frequency emissions to the extent that such facilities comply with the Commission's regulations concerning such emissions." The FCC has displaced local government control

This language effectively prohibits discussion of environmental concerns or health and safety concerns in the placement of cell towers. This is despite growing awareness and scientific confirmation of both environmental and health effects from exposure to cell tower radiation and all radiofrequency wireless devices where they have introduced the new technology (Physicians for Safe Technology, Telecommunications Act of 1996). People are upset about thousands of pieces of equipment the size of a pizza box or briefcase being deployed to "vertical infrastructures"—municipal utility poles, street lights, and even traffic lights, just outside their homes.

An opinion piece in the LA Times noted that locals are crying foul, calling it an audacious power grab and the equivalent of a gift of public funds to billion-dollar telecommunication companies. These local entities should be able to retain some authority to push back on proposed deployments. After all, cell phone companies' goals are for making profits and serving customers, not making a city look nice. (LA Times Editorial Board 2017) www.latimes.com/opinion/editorials/la-ed-power-pole-grab-20170705-story.html)

Local lawmakers must be wary and show interest in protecting the rights of communities to govern the use of their infrastructure rather than letting telecommunication companies make those decisions for them.

A US court has found the FCC erred in trying to exempt 5G cell sites from environmental impact and historic preservation reviews in a case brought on by environmental and tribal groups. However, the judges declined to address the FCC's move forcing cities to approve cell sites on expedited time frames, or requirements aimed at reducing the cost of permitting. Those are the subject of a separate case in the US Court of Appeals for the Ninth Circuit. (Fung, CNN Business: Court deals blow to FCC's bid to speed 5G rollout).

The FCC and other agencies

On June 20, 2016, former FCC Chairman Tom Wheeler gave an overview of the FCC's philosophy of the roll-out of 5G at a National Press Club Luncheon. "Unlike some countries, we do not believe that we should spend the next couple of years studying what 5G should be or how it should operate and how to allocate spectrum, based on those assumptions," he told the group. (Wheeler 2016) https://www.press.org/sites/default/files/20160620_wheeler.pdf

Wheeler continued that while "the European Union has put up 700 million Euros to do 5G research. . .we think that's the wrong way to go. We think that making the Spectrum available, and standing out of the way of technology development, is far better. . ." than sitting around waiting to decide what it's going to do and micromanaging the technology process.

Stepping back from regulatory responsibilities, he explained, "We won't wait for the standards to be first developed in the sometimes arduous standards setting process or in government led activity. Instead we will [make] ample spectrum available and then rely on a private sector led process for producing technical standards best suited for those frequencies and use cases."

Current Federal Communications Commission Chairman, Ajit Pai, has said the agency places a "high priority on the safety of wireless services and devices" He said the agency's guidelines for radio frequency (RF) exposure are derived from guidance from the EPA, as well as the IEEE, the International Commission on Non-ionizing Radiation Protection, and the National Council on Radiation Protection and Measurements. https://www.cnet.com/news/5g-phones-and-your-health-what-you-need-to-know/

"The FCC relies on the expertise of health and safety agencies and organizations with respect to appropriate levels of RF exposure," Pai said. "These institutions have extensive experience and knowledge in RF-related issues and have spent a considerable amount of time evaluating published scientific studies that can inform appropriate exposure limits." https://www.fiercewireless.com/5g/fcc-responds-to-5g-health-concerns-as-politicians-push-for-more-information

The FDA, which is also responsible for ensuring the safety of cellphones, has said that it "continues to believe that the current safety limits for cellphone radio frequency energy exposure remain acceptable for protecting the public health." https://www.cnet.com/news/5g-phones-and-your-health-what-you-need-to-know/

A major problem is that guidelines for non-ionizing radiation exposure were developed over two decades ago, long before the development of 5G, and are based on heating of tissues over short exposure periods (FCC 1997b, 2015: FCC 2013).

With the passage of the Federal Telecommunication Act of 1996, responsibility for safety of non-ionizing radiation was passed from the EPA to the FCC. At the time, the EPA was preparing recommendations for long term exposure which were not included in the FCC guidelines (EPA 1981; EPA 1992). At a 1993 scientific conference sponsored by the US EPA Office of Air and Radiation, and the Office of Research and Development, the EPA discussed its concerns about public radio frequency (RF) exposure and the need for additional research. The report noted health issues that remained unsolved including "potential effects of long term, low level exposure, and biophysical mechanisms" (EPA 1993).

The Food and Drug Administration (FDA) and the FCC say there's nothing to be worried about. They claim there are studies that haven't found a link between radio frequency signals from cellphones or cell towers and disease. But there are studies that have. In fact, the World Health Organization has found that cellphones might cause some brain cancers, leaving open the possibility that a link exists between cancer and cellphone radiation.

Unfortunately, the FCC has not updated the cellphone safety standards since 1996. Critics say the cellphone safety standards should be reviewed based on the latest wireless technology. In March 2019, the engineering group, the Institute of Electrical and Electronics Engineers, (IEEE) recommended the safety levels remain about the same as they have been since 1996.

California Congresswoman Anna Eshoo introduced Federal bill HR 530 in the House of Representatives, with 53 co-sponsors, to invalidate the Federal Communications Commission's (FCC) September 26, 2018 ruling to accelerate the deployment of 5G small cells throughout the US in early 2019. The bill is titled "To provide that certain actions by the Federal Communications Commission shall have no force or effect." The bill was referred to the Subcommittee for Communications and Technology, where it currently sits without action for over a year.

Senator Diane Feinstein, with eight co-sponsors, introduced a bill to the Senate for the same purpose in June

2019. The bill, "To provide that certain regulatory actions by the Federal Communications Commission shall have no force or effect," was referred to the Committee of Commerce, Science, and Transportation, where it also sits without further action.

Members of Congress Defozio and Suozzi co-authored a letter to the FCC in September of 2019, accusing them of a lack of transparency and undermining the public's trust in the FCC and the Federal government. They noted the Chicago Tribune's expose' which found that some popular cell phone models emit higher RF radiation than are allowed under FCC regulations. For example, the Samsung Galaxy S8 tested at more than five times the FCC's exposure limits. The FCC agreed to additional testing of cell phones. (Defozio and Suozzi September 2019).

Deployment of 5G

Many local governments are trying to stop the 5G rollout, insisting that wireless companies prove the technology isn't harmful to people. But the Telecom Act of 1996 prohibits local governments from using health or safety concerns as a reason to block cellphone deployment. Rather than answer to the requirements of local utility companies and local governments, the telecommunications industry need only show they are in compliance with FCC set standards. The FCC has set the health and safety requirements, as well as made rules about the amount that can be charged for using a utility pole for an antenna(s), and produced a "shot clock" to enable speedy passage of the application for use.

Kenneth Foster, professor of Bioengineering, University of Pennsylvania sits on the IEEE's standards committee for setting radio frequency exposure limits. He argues the reason is that the regulatory system requires that manufacturers of RF-emitting equipment verify compliance with FCC safety limits, rather than directly conducting toxicity studies. The industry is basically self-governing.

In the first six months of 2019 alone, the FCC approved more than 21,000 RF-emitting devices across the entire frequency range. None of these have to be subjected to comprehensive toxicity testing. Rather, all they have to do is to show they are in compliance with FCC safety limits together with other regulations. And yet the Chicago Tribune's expose' showed some cell phones emit higher

RF radiation than are allowed under the FCC's safety standards.

Many experts say more quality research is needed based on countless studies and protest letters and calls for a moratorium which show a link with cancer. Despite the scientific evidence, the FCC in the US is accelerating the deployment of wireless antenna infrastructure (FCC Fact Sheet, 2018).

Dr. Dariusz Leszczynski, of Helsinki, Finland, points out that there is a lot of confusion around 5G; it is "being developed as well as deployed at the same time. Even the technical standards dealing with the 5G are not all ready yet." (Leszczynski 2019).

The FCC standards for 5G gigahertz wavelengths urgently need reevaluation. The short millimeter waves for 5G technology bring along new challenges for public safety. To improve the performance of 5G and IoT (Internet of Things) devices, pulsed phased antenna arrays will be used in cell towers and wireless devices.

There are also additional concerns specific to 5G due to the super high-frequency millimeter wavelengths used. Because signals transmitted over millimeter waves are limited in range and can't penetrate obstacles like walls or even leaves on trees, networks using these frequencies will require radios on every city block versus 4G gear that transmits signals over miles. Not only will there be more 5G antennas transmitting signals, but the antennas will have to be closer to you.

Thermal and biological effects of 5G

At this point, according to the IEEE, heat is still the only standard of harm considered in 5G applications, not biological effects. Current FCC policy does not consider non-thermal biological and physiological radio frequency effects on living organisms, despite abundant literature for 2G, 3G, and 4G systems.

Sage and Carpenter, among others, note that for adequate public health protection, a biological safety standard (in addition to the heat safety standard) is needed that considers current research indicating cellular harm, long term effects of constant exposure and effects on vulnerable populations (Sage and Carpenter 2012; Blank 2015). FCC recommendations have not been updated to include current literature on cellular affects at levels below FCC guidelines or effects of long-term exposure.

The current International Commission on Non-ionizing Radiation Protection/World Health Organization (ICNIRP/WHO) guidelines for EMF are based on the obsolete hypothesis that the critical effect of RF-EMF exposure relevant to human health and safety is heating of exposed tissue. However, scientists have proven that many different kinds of illnesses and harms are caused without heating ("non-thermal effect") at radiation levels well below ICNIRP guidelines (BioInitiative 2012 Report).

It has long been believed that non-ionizing RF radiation could not be harmful to living organisms because RF radiation did not break chemical bonds or have enough energy to remove an electron from (ionize) an atom or molecule. It was felt radio frequency radiation damaged tissue only through a heating or burning mechanism.

However, a review of the research in 2010 noted that "A large number of cellular studies have indicated the millimeter wavelengths (MMW) may alter structural and functional properties of membranes." Exposure to MMWs may affect the plasma membrane either by modifying ion channel activity or by modifying the phospholipid bilayer. Water molecules also seem to play a role in these effects. Skin nerve endings are a likely target of MMWs and the possible starting point of numerous biological effects. MMWs may activate the immune system through stimulation of the peripheral neural system (Ramundo-Orlando 2010).

We are now learning that microwave radio frequency electromagnet radiation (EMR) from wireless devices acts as an environmental stressor with direct oxidative toxic effects on cellular processes that are not related to heat or to ionization. The effect of radio frequency EMR is indirect, inducing biochemical changes in cellular structures and their membranes.

An older Russian paper, "Biological Effects of Millimeter Wavelengths" by Zalyuboxskaya (1977) was declassified by the CIA in 2012. This paper describes the research on both humans and animals showing a myriad of adverse effects of millimeter wavelengths. The author notes that millimeter wave technology had been used for years without any studies on biological effects. The researchers found that "millimeter waves caused changes in the body manifested in structural alterations in the skin and internal organs, qualitative and quantitative changes in the blood and bone marrow composition, and changes in the conditioned reflex activity, tissue respiration . . . and nuclear metabolism. The degree of unfavorable effect of millimeter waves depended on the duration of radiation and individual characteristics of the organism."

The author confirmed that millimeter waves do not penetrate skin but act on nerve receptors in the skin to cause such diverse biological and metabolic effects as a reduction in hemoglobin and erythrocytes, higher blood cortisol levels, adrenal stimulation, mitochondrial dysfunction and suppression of the central nervous system with notable changes in liver, kidneys, heart, and brain.

The 2018 European Commission Scientific Committee on Health, Environmental and Emerging Risks (SCHEER) lists 5G electromagnet radiation as an emerging concern. They state, "The lack of clear evidence to inform the development of exposure guidelines to 5G technology leaves open the possibility of unintended biological consequences." (SCHEER 2018).

Betzala (2018) demonstrated that the sweat glands which are coiled structures in the upper layers of the skin can act as an antenna receiver for 5G sub-THz band wavelengths. The presence of sweat glands in the skin significantly increase the specific absorption rate (SAR), or heat absorption for millimeter radio frequency radiation. This varies depending on factors such as perspiration and stress levels.

Nielson (2019) states, "Weak radio frequency (RF) magnetic fields in the MHz-range was shown to influence the concentrations of reactive oxygen species (ROS) in living cells (Nielson 2019, Usselman 2016, 2014l Castello 2014; Naarala 2017)." Oxidative stress denotes high levels of ROS that incur damage to DNA, protein or lipids.

Belpomme (2018) in "Thermal and non-thermal health effects of low intensity non-ionizing radiation: An international perspective," finds that exposure to low frequency and radiofrequency electromagnetic fields at low intensities poses a significant health hazard that has not been adequately addressed by national and international organizations such as the World Health Organization

A growing body of scientific literature documents evidence of non-thermal cellular damage from non-ionizing wireless radiation used in telecommunications. This RF EMR has been shown to use an array of adverse effects on DNA integrity, cellular membranes, gene expression, protein synthesis, neuronal function, the blood brain carrier, melatonin production, sperm damage and immune dysfunction (Dasdag 2015a, 2015b; La Vignere 2012, Levine 2017).

Human health effects associated with wireless radiation include infertility, neurodegenerative changes, and brain cancer (Wyde 2016; IARC 2011; Sage and Carpenter 2012; Kim 2017; Kesari 2011, 2012a, 2012b; Zhang 2016; Agarwal 2011, 2008; Al-Quzwini 201; Banik 2003; Consales 2012; D'Andrea and Chalfin 2000; Desai 2009; Prasad 2017).

In addition, electrosensitivity to wireless and electrical devices is being increasingly recognized by scientists and physicians (Hojo 2016; Belpomme 2015). There is also growing evidence of harm to trees, wildlife and other biosystems (Sivani and Sudarsanam 2013).

A well-studied potential mechanism of harm from radiofrequency radiation is one of cellular oxidation. Healthy biological systems require a balance of oxidation and anti-oxidation to fight infection and prevent disease. A review of the literature by Yakymenko (2016) confirmed that in 93 of 100 studies, non-ionizing radio frequency radiation caused a cellular stress response with excessive reactive oxygen species (ROS). He concluded, "Oxidative stress induced by RFR exposure should be recognized as one of the primary mechanisms of the biological activity of this kind of radiation."

In humans, the penetration depth of more than 90% of the transmitted power is absorbed in the epidermal and dermal layers (Wu 2015a). Because the depth is so superficial, higher heating occurs more quickly with less dissipation. Many biological responses to MMW irradiation can be initiated with the skin (Isaac 2012; Ziskin 2013; Gandhi and Riazi 1986).

In a review article, Pakhomov (1998) looked at the biological effects of millimeter wavelengths (MMW) of 5G. He examined dozens of studies and cites research demonstrating profound effects of MMW on all biological systems including cell, bacteria, yeast, animals, and humans. Some effects were clearly thermal, however, many of the studies showed non-thermal biological effects at low intensities.

In 2018, the National Toxicology Program (NTP), an inter-agency program run by the US Department of Health and Human Services, published final results of its decade-long study on rats. The study found a link between exposure to high levels of 2G and 3G cellphone radiation and cancerous heart and brain tumors in male rats. (NTP 2018).

The concerns about 5G are similar to the concerns about 2G, 3G and 4G. While 5G in the US is expected to use some of the same frequency bands that previous generations of wireless have used, including low band 600 MHz frequencies as well as midband spectrum in the 2.g GHz, 3 GHz, and 3.7-4.2 GHz bands, operators such as AT&T and Verizon are also targeting higher frequency bands for 5G. The FCC has already auctioned off airwaves in the 24 GHz and 29 GHz bands. And it will soon auction off licenses in the 27 GHz, 30 GHz, and 47 GHz bands.

It's this "high band" spectrum that's of most concern, because it'll require denser radio deployments. And there's also less research on the effects of radiation at these higher frequency bands.

Martin Pall, a professor emeritus of Biochemistry and Basic Medical Sciences at Washington State University, says the evidence is clear that cellphone radiation is dangerous. He says results from existing studies show clear lines between cellphone radiation and a wide range of medical maladies from cancer, to infertility, to depression.

"What we're doing is destroying our health." Pall said in an interview. He added that if 5G deployments aren't stopped, "We [as a society] are playing around with our very survival." https://www.cnet.com/news/5g-phones-and-your-health-what-you-need-to-know/

Kenneth Foster, Professor of Bioengineering at the University of Pennsylvania, sits on the IEEE's standards committee for setting radio frequency exposure limits. He acknowledges that unlike at 3G and 4G radiation levels, which have been studied for at least two decades, there isn't as much research on the biological effects of using millimeter wavelengths for 5G service.

Since widespread cellphone use is relatively recent, we could still be a number of years away before we see an epidemic of cancer due to cellphone radiation, according to Dr. Jonathan Samet, who chaired the WHO's 2011 committee on cellphone radiation. It's still too early to know based on population studies if cellphone radiation causes tumor growth in humans. He points out it took at least 20 to 25 years after cigarettes began being mass produced for epidemiologists to notice the link between lung cancer and smoking tobacco.

Mortazavi (2019) reports the radio frequency electromagnet fields (RF-EMFs) produced by widely used mobile phones are classified as possibly carcinogenic to humans by WHO (World Health Organization), International Agency for Research on Cancer (IARC). The current data on the relationship between exposure to RF-EMFs plays a basic role in RF-induced carcinogens.

Increased risk of brain cancer after heavy or long-term mobile phone use has been reported by different researchers (INTERPHONE 2010; Roosli 2013; Coureau 2014; Carlberg 2015). It has also been reported that exposure to RF-EMFs may cause stimulatory or inhibitory effect on the proliferation and differentiation of stem cells (Eghlidospour 2015).

Cell phone industry and FCC conflict of interest

There are many lessons we have not learned with the introduction of novel substances, which later became precarious environmental pollutants by not heeding warning signs from scientists (Gee 2009).

Russell (2018a) notes that industry continues to state that the weight of evidence regarding harm from RF-EMR is inconclusive. Studies that review the sources of funding and scientific bias regarding cell phones and brain cancer indicate otherwise. Huss (2007) performed a systematic review regarding the association of cell phone use and brain tumors in relation to funding. He found that industry studies showed a positive association 33% of the time, whereas non industry studies showed an 82% association.

Climate change, fracking, toxic emissions and microwave radiation from wireless devices all have something in common with smoking. There is much denial and confusion about health and environmental risks, along with industry insistence for absolute proof before regulatory action occurs (Frentzel-Beyme 1994; Michaels 2008).

An analysis of the INTERPHONE study by Morgan (2009) noted eleven design flaws, including 1) selection bias, 2) insufficient latency time, 3) definition of 'regular' cellphone user, 4) exclusion of young adults and children, 5) no consideration for cell phone exposure in rural areas where they would be radiating at higher power levels, 6) exposure to other transmitting sources are excluded, 7) exclusion of brain tumor types, 8) recall accuracy of cellphone use, and 9) funding bias.

A study by Kostoff, PhD, School of Public Policy, Georgia Institute of Technology, "Adverse Effects of Wireless Radiation" includes a section on why not all studies of wireless radiation have shown adverse effects. These include 1) There may be "windows" where adverse effects occur, and operation outside these "windows" would show a) no effects, or b) hormetic effects (dose-response), or c) therapeutic effects. 2) Research quality could be poor, and adverse effects were overlooked. Or, 3) the research team could have had a preconceived agenda, where finding no adverse effects from wireless radiation was an objective (Kostoff 2019).

Kostoff bluntly states, "For example, studies have shown that industry-funded research of wireless radiation adverse health effects are far more likely to show no effects than funding from non-industry sources (Huss 2007; Slesin 2006; Carpenter 2019). Unfortunately, given the strong dependence of the civilian and military economies on wireless radiation, incentives for identifying adverse health effects from wireless radiation are minimal and disincentives are many.

Kostoff continues, "Even the Gold Standard for research credibility—independent replication of research results—is questionable in politically, commercially, and militarily sensitive areas like wireless radiation safety." Suppose there are two research groups, funded by the same government agency, who both arrive at the same conclusion that just coincidentally coincides with what the government sponsor wanted. Would this be considered independent? Or, these two research groups received funding from different agencies of the government. Review articles tend to treat either of these cases as independent for statistical purposes, and don't make the distinction as long as the validation doesn't arise from within the performer group.

Even reporting of conflict-of-interest on wireless radiation research papers or evaluation panels leaves much to be desired. Currently, potential conflicts of interest of the research performers are identified by listing the funding sources in the published papers, or other formal documented evidence of conflicts of interest. However, there are many potential conflicts of interest that may not be as formal, but could be at least as influential as the formal conflicts in determining the outcome of the research or proposal. Kostoff points out that to ascertain these other less formal conflicts of interest would require vetting:

- any elements of the researchers'/evaluators' investment portfolio that would profit from operation and expansion of the mobile telecommunications network, including impacts on related industries;
- any elements of their present business endeavors that would profit from operation and expansion of this network, including impacts on related industries;
- any elements of present or future pensions that would profit from operation and expansion of this network, including impacts on related industries;
- any proposals or future employment offers in the pipeline or being considered that would profit from operation and expansion of this network, including impacts on related industries;
- any other listing or potential conflicts of interest by which they could profit from operation and expansion of the mobile telecommunications network, including impacts on related industries.

Kostoff concludes, "Wireless radiation offers the promise of improved remote sensing, improved communications and data transfer, and improved connectivity. Unfortunately, there is a large body of data from laboratory and epidemiological studies showing the previous generations of wireless networking technology have significant adverse health impacts. Much of this data was obtained under conditions not reflective of the real-world. When real-world considerations are added, such as 1) including the information content of signals along with 2) the carrier frequencies, and 3) including other toxic stimuli in combination with the wireless radiation, the adverse effects are increased substantially. Superimposing 5G radiation on an already imbedded toxic wireless radiation environment will exacerbate the adverse health effects shown to exist. Far more research and testing of potential 5G health effects is required before further rollout can be justified."

But the power of the cellular industry may be larger than scientific proof. In "Captured Agency: How the Federal Communications Commission is Dominated by the Industries it Presumably Regulates," Norm Alster points out the FCC is a "captured agency." Captured agencies are essentially controlled by the industries they are supposed to regulate, and it results in the wireless industry pretty much getting what it wants. The FCC simply echoes the lobby points of major technology interests.

The Cellular Telecommunications Industry Association (CTIA) and the National Cable and Telecommunications Association (NCTA) have annually been among Washington's top lobbying spenders. Members of congressional oversight committees are prime targets of the industry. However, money, thru lobbying, is just one part of it.

The industry is allowed access in a variety of ways to shape the FCCs policies, often at the expense of fundamental public interests. As a result, consumer safety, health, and privacy, along with consumer wallets, have all been overlooked, sacrificed, or raided due to unchecked industry influence. Industry extends beyond Congress and regulators to basic scientific research. In a parallel to the hardball tactics of the tobacco industry, the wireless industry has backed up its economic and political power by stonewalling on public relations and bullying potential threats into submission with its huge standing army of lawyers.

There is also the free flow of executive leadership between the FCC and the industries it presumably oversees. Members of Congress and employees become employed by the wireless industry, and wireless executives are appointed to leadership positions in the FCC. Staffers who cooperate with the wireless industry often leave their government positions to become lobbyists themselves.

One of the biggest accomplishments of the wireless industry in capturing the FCC has been the silencing of the public. As a section of the Communications Act of 1996 provides: "No State or local government or instrumentality thereof may regulate the placement, construction and modification of personal wireless service facilities on the basis of the environmental effects of radio frequency emissions to the extent that such facilities comply with the Commission's regulations concerning such emissions."

The FCC preempted local zoning authority, along with the public's right to guard its own safety and health. This has resulted in industry having a free hand in installing more than 300,000 sites. Congress and the FCC have extended the wireless industry carte blanche to build out infrastructure no matter the consequences to local communities. The FCC in 1997 sent the message it has implicitly endorsed and conveyed ever since--Study health effects all you want. It doesn't matter what you find. The build-out of wireless cannot be blocked or slowed by health issues (FCC 1997a).

Since the impact of RF diminishes with distance, industry advocates and many scientists dismiss the possibility that such structures have health risks. But there is a body of evidence that suggests exposure to even low emission levels at typical cellular frequencies between 300 MHz and 3 GHz can have a wide range of negative effects. The FCC has turned a blind eye to these reports, citing instead other studies—often industry-funded—that fail to establish health effects.

Need for further research

MMW research is sparse; there is a need for independent testing for both long-term and short-term effects. There are EMF studies which show negative effects that are not taken seriously by the institutions which make public policy on the safety of the American people. There is an extensive need for additional research.

Russell (2018a) points out that considering current peer reviewed science, predicable harm to life forms within the mixed frequency networks with negative consequence appears likely over time. For electrosensitive individuals, it will add to their physical symptoms and isolation with significant reduction in non-exposed safe havens. There is an urgent need for independent studies to guide development of effective public health standards and policies.

In "Adverse Effects of Wireless Radiation," Ronald N Kostoff, Ph.D. points out that "almost all of the laboratory tests that have been performed are flawed with respect to showing the full adverse impact of the wireless radiation. Either 1) non-inclusion of signal information, or 2) using single stressors only, tends to underestimate the seriousness of the adverse effects from non-ionizing radiation."

He explains the typical incoming EMF signal for most laboratory tests performed in the past consisted of the single carrier wave frequency; the lower frequency superimposed signal containing the information was not always included. This omission may be important because as Panagopoulos states, except for the RF/microwave carrier frequency, Extremely Low Frequencies – ELFs (0-3000Hz) are always present in all telecommunication EMFs in the form of pulsing and modulation. There is significant evidence indicating that the effects of telecommunication EMFs on living organisms are mainly due to the included ELFs. . .While ~50% of the studies employing simulated exposures do not find any effects, studies employing real-life exposures from commercially available devices display an almost 100% consistency in showing adverse effects (Panogopoulos 2019)

These effects may be exacerbated further with 5G: "with every new generation of telecommunication devices . . . .the amount of information transmitted each moment . . . .is increased, resulting in higher variability and complexity of the signals with the living cells/organisms even more unable to adapt (Panogopoulos 2019)."

Kostoff further explains these tests typically involve one stressor (toxic stimulus) and are performed under pristine conditions. This does not represent real-life exposures, where humans are exposed to multiple toxic stimuli, in parallel or over time. In only about 5% of the cases reported in the literature, a second stressor (biological or chemical toxic stimuli) was added, to ascertain whether additive, synergistic, or antagonistic effects were generated by the combination (Kostoff and Lau 2017; Juutilainin 2008; 2006).

It's important to have combination experiments because when other toxic stimuli are considered in combination with non-ionizinng EMF radiation, the synergies tend to enhance the adverse effects of each stimulus in isolation. In other words, Kostoff points out, combined exposure to toxic stimuli and non-ionizing EMF radiation translates into much lower levels of tolerance for each toxic stimulus in the combination relative to its exposure levels that produce adverse effects in isolation.

So, the "exposure limits for non-ionizing EMF radiation when examined in combination with other potentially toxic stimuli would be far lower for safety purposes than those derived from non-ionizing EMF radiation exposures in isolation (Kostoff 2019)." Therefore, the results reported in the biomedical literature should be viewed as extremely conservative and the very low "floor" of the seriousness of the adverse effects, not the "ceiling".

The American Cancer Society generally finds most studies published so far have not found a link between cell phone use and the development of tumors. However, they point out on their website that these studies have had some important limitations that make them unlikely to end the controversy about whether cell phone use affects cancer risk.

They write, "First, studies have not yet been able to follow people for very long periods of time. When tumors form after a known cancer-causing exposure, it often takes decades for them to develop. Because cell phones have been in widespread use for only about 20 years in most countries, it is not possible to rule out future health effects that have not yet appeared (ACS 2019)."

"Second, cell phone usage is constantly changing. People are using their cell phones much more than they were even 10 years ago, and the phones themselves are very different from what was used in the past. This makes it hard to know if the results of studies looking at cell phone use in years past would still apply today."

"Third, most of the studies published so far have focused on adults, rather than children. (One case-control study looking at children and teens did not find a significant link to brain tumors, but the small size of the study limited its power to detect modest risks.) Cell phone use is now widespread even among younger children. It is possible that if there are health effects, they might be more pronounced in children because their bodies might be more sensitive to RF energy. Another concern is that children's lifetime exposure to the energy from cell phones will be greater than adults', who started using them at a later age."

Do electromagnet fields cause cancer?

It has been estimated there were almost 5 billion mobile phones worldwide in 2019. This is up from one billion reported to exist in 2015. The increasing use of mobile phones has raised concerns of health risk and especially for intracranial tumors, since the brain is the nearest organ that is in close contact with the radio frequency electromagnet fields emitted by mobile phones. An even greater increased risk has been suggested for children due to thinner skull, smaller head and increased brain conductivity (Wiart 2008).

It was previously believed that as RF-EMFs do not have enough energy to remove electrons (producing an ion pair), they are unable to cause cancer. However, considering the evidence of free-radical damage which has been confirmed in studies on humans, animals, plants and microorganisms, some scientists now believe that EMF-induced oxidative stress can cause damage to cellular targets such as DNA and trigger processes which lead to cancer (Havas 2016).

Here is the conundrum: there are both case control (Shrestha 2015) and cohort (Johansen 2001; Schuz 2006; Frei 2011), studies which could not link the RF-EMF exposure to cancer. But there are more recent studies that do show a link.

The INTERPHONE Study Group previously investigated the relationship between brain tumor (glioma and meningioma) risk and mobile phone use (Interphone Study Group 2011). In this study, no elevated odds ratio

(OR) for glioma or meningioma was observed more than ten years after first phone use, except in cases of heavy phone use. The study was coordinated by the International Agency for Research on Cancer (IARC), a part of WHO, the World Health Organization. The fact is, however, it did find an association for heavy phone use.

Myung (2009) performed a meta-analysis and found that there was a small but significant elevation in brain tumors with long term cell phone use when high quality studies were examined. He noted Hardell's research [which found an association with cancer] to be more robust, as "all of the studies by Hardell, used blinding to the status of patient cases or controls at the interview and were categorized as having a high methodology quality when assessed based on the NOS, whereas most of the INTERPHONE-related studies and studies by other groups did not use blinding and were thus categorized as having low methodiologic quality."

In Australia, a country in which mobile phone use was started in 1987, one study reported that despite the steep increase in mobile phone use, no compatible increase in brain cancer incidence has been found (Chapman 2016).

But Mortazavi recently reviewed currently published papers claiming no link between exposure to RF and brain cancer. They found that in many cases there were large errors and/or major shortcomings in these papers (Mortazavi SAR 2017). For example, in one of the papers reviewed by this research group ("Analysis of Mobile Phone Use Among Young Patients with Brain Tumors in Japan," Sato 2017), a 400% difference in brain tumors was masked by statistics (Mortazavi SMJ 2017). Considering the controversies existing today, they reported these controversies may be caused by several key parameters, especially the large difference in the magnitude of exposure to RF-EMFs in different studies.

On the other hand, the recent $25 Million large-scale animal study conducted by the U.S. National Toxicology Program (NTP) showed statistically significant increases in cancer in rodents exposed to GSM or CDMA signals for two-years (Wyde 2018). This study showed that malignant gliomas in the brain and schwannomas of the heart could be linked to mobile phone exposures.

Hardell and Carlberg in 2015 performed a pooled analysis of increased risk of two case-control studies on malignant brain tumors and showed that mobile phone use was linked to increased risk of glioma (Hardell 2015).

Moreover, based on the findings of a case-control study on brain tumors, a significant association was found between mobile and cordless phone use and malignant brain tumors. The authors reported that their findings support the hypothesis that exposure to RF-EMFs generated by wireless phones can play a key role in the initiation and promotion phases of carcinogenesis. (Hardell 2013b).

Bortkiewicz recenty performed a meta-analysis of twenty four studies, for a total of 26,846 cases, 50,013 controls, and reported their findings supported the hypothesis that long term use of mobile phones is linked to increased risk of intracranial tumors (Bortkiewicz , 2017). They reported that mobile phone use over ten years was significantly associated with higher risk of intracranial tumors (all types.)

Another meta-analysis performed by Wang and Guo showed a significant association between mobile phone use greater than five years and the risk of glioma (Wang 2016).

Some of the current studies have introduced the improvements in diagnostic procedures as the reason for increased cancer incidence (Chapman 2016). However, other researchers reported the increased cancer rate cannot be attributed to better diagnostic procedures and suggested that the role of exposure to both ionizing (e.g., rapidly increased number of CT scans) and non-ionizing radiation should be further studied (Carlberg 2016).

Although the current controversy about the role of exposure to RF-EMFs on cancer incidence may be due to several parameters, it seems that the level of exposure plays a basic role in this issue. In a study performed in France on a possible relationship between mobile phone exposure and primary central nervous system tumors (gliomas and meningiomas) in adults, it was found that heavy mobile phone use could be linked to brain tumors (Coureau 2014).

A case-control study in Finland found no excess risk associated with self-reported short term and medium-term use of mobile phones. However, the authors claimed there are uncertainties for long term use in that only a small proportion of participants were long term users in their study (Shrestha 2015).

Prasad has also provided evidence which proves an association between mobile phone use and brain tumors especially in people who used their mobile phones for more than ten years (Prasad 2017).

Yakymenko reviewed the published data on carcinogenic effects of long-term exposure to low intensity microwave radiation (Yakymenko 2011). It was reported the carcinogenic effect of radio frequency radiation should typically be manifested after long term exposures, e.g., durations more than ten years. In addition, Alexiou and Sioka have reported that there was some evidence to suggest a connection between heavy mobile phone use and increased risk for brain tumor occurrence especially for gliomas (Alexiou 2015).

In 2011, the WHO International Agency for Research on Cancer categorized radio frequency electromagnetic fields (RF) from mobile phones, and from other devices, as a Group 2B, a possible human carcinogen. This list includes DDT, lead, chloroform, insecticides and chemicals. (WHO 2011).

Investigation of the effects which RF may produce in cellular level *in vivo* and *in vitro* revealed increased risk of cell death and cancer development in mice. The potential effect of RF to germ cells is worrisome since it can be transmitted to subsequent generations (Liu 2013).

Call for a Moratorium

Sen Richard Blumenthal, a Democrat from Connecticut and several other Democrats in the House of Representatives are demanding the FCC demonstrate that 5G is safe. (Blumenthal 2019) During a February, 2019 Senate Commerce, Science, and Transportation Committee hearing on the future of 5G wireless technology and their impact on the American people and economy, Senator Blumenthal raised concerns with the lack of any scientific research and data on the technology potential health risks.

Blumenthal blasted the FCC and the FDA—government agencies jointly responsible for ensuring that cellphone technologies are safe to use—for failing to conduct any research into the safety of 5G technology, and instead, engaging in bureaucratic finger-pointing and deferring to industry. The wireless industry representatives conceded they have not supported independent research on the safety of 5G technology and the potential links between radiofrequency and cancer.

The International EMF Scientist Appeal was launched on September 13, 2017 and by December 18, 2019, had been signed by 268 scientists who've previously published peer reviewed research on the health effects of EMF. It is a clarion call for the UN, as well as its sub-organizations like the WHO, to enact stricter regulations on the amount of EMF people are legally allowed to be exposed to.

They begin by saying, "We the undersigned, scientists and doctors, recommend a Moratorium on the roll-out of the fifth generation, 5G, for telecommunication until potential hazards for human health and the environment have been fully investigated by scientists independent from industry. 5G will substantially increase exposure to radio frequency electromagnet fields "RF-EMF" on top of the 2G, 3G, 4G, Wi-Fi, etc., for telecommunications already in place. RF-EMF has been proven to be harmful for humans and the environment."

The scientists and doctors refer to the fact that "numerous recent scientific publications have shown that EMF affects living organisms at levels well below most international and national guidelines." Effects include increased cancer risk, cellular stress, increase in harmful free radicals, genetic damages, structural and functional changes of the reproductive system, learning and memory deficits, neurological disorders, and negative impacts on general well-being in humans.

On June 9, 2017, scientists with the International EMF Scientist Appeal submitted a letter of comment to the Federal Communications Commission (FCC) in opposition to allowing streamlined approval of 5G infrastructure to be built on existing utility poles, in greater number than current cellular antennae.

Earlier, in 2016, Dr. Yael Stein of Jerusalem's Hebrew University sent a letter to the FCC Commission (https://ehtrust.org/letter-fcc-dr-yael-stein-md-opposition-5g-spectrum-frontiers/) and the U.S. Senate Committee on Health, Education, Labor and Pensions, and the U.S. Senate Committee on Commerce, Science and Transportation, outlining the effect of 5G radiation on human skin.

According to Dr. Stein, over ninety percent of microwave radiation is absorbed by the epidermis and dermis layers of human skin, essentially making it an absorbing sponge for microwave radiation. Also, the sweat ducts in our skin's upper layer act like helical antennas, which are antennas specially designed to respond to electromagnet fields. In essence, our bodies will conduct 5G radiation, and how this will affect babies, pregnant women, and the elderly is not known.

Damage goes well beyond the human race, as there is growing evidence of harmful effects (Nittby 2011) to both plants (Waldman-Selsam 2016) and animals (Balmori 2009).

The Danish Institute for Public Health and the council for Health-Safe Telecommunications has prepared a legal document related to the broad harm from 5G as well as other wireless technologies. They state, "The legal opinion is based on the rules of law in the European Convention on Human Rights, the UN Convention on the Rights of the Child, the EU directive on the conservation of natural habitats and of wild fauna and flora, the EU directive on the conservation of wild birds, on the precautionary principle, as well as on the Bern- and Bonn-conventions on the protection of animals and plants" (Jensen 2019).

Lawmakers and policy makers throughout the world are starting to put on the brakes. In April 2019, the Belgian government halted a 5G test in Brussels over concerns that radiation from the base stations could be harmful.

In the Netherlands, their Parliament asked for an independent investigation on 5G health risks prior to it being installed on a large scale. The municipality of Rome has put forth a resolution asking the mayor stop the 5G trial and not to raise the limit values in the threshold of electromagnetic radiation, avoiding the positioning of millimeter microwave antennas on homes, schools, day centers, recreation centers, street lamps and more. German petitioners have requested the German

Parliament suspend the award of 5G frequencies based on scientifically justified doubts about the safety of this technology. And there is debate in England on health-related effects of electromagnetic fields and 5G led by MP Tonia Antoniazzi.

5G has been launched in 102 locations in Switzerland. People living in Geneva after the 5G rollout reported alarming details of illness in an article "With 5G, We Feel Like Guinea Pigs, posted July 18, 2019 in L'Illustre magazine.

In the article, people living with 5G technology were living heathy lives. Since the installation of 5G antennas, they have reported having trouble falling asleep, not feeling well, headache, fatigue, sinus conditions, chest pain, pain on the left side the head and on the back of the skull, and tinnitus. When they complained to Swisscom, who installed the technology, they were told that "tests had taken place and that everything was in order."

The residents report they feel they are not being heard. "As soon as we come up against the financial interests of these people, they are in total denial." one laments, continuing, "If there is indeed a multiplication of leukemias or brain cancers, it will take years to be noticed." There is little question of moving when there will be antennas everywhere soon. (Technology: 5G Telecommunications Science. 12/02/19 https://mdsafetech.org/5g-telecommunications-science/)

In December, 2019, a broad coalition of scientists, doctors and advocates sent a National 5G Resolution letter to President Trump demanding a moratorium on 5G until potential hazards for human health and the environment have been fully investigated by scientists independent from the telecom industry (National 5G Resolution letter: https://entrust.org/usa-national-5g-resolution).

The letter references the published scientific studies demonstrating harm to human health, bees, trees, and the environment from current wireless technology and posits that 5G will both increase exposure and add in new technology never safety tested for long-term exposure.

Ways to protect yourself

Whenever your cell phone is turned on, it is emitting radiation signals, and has the potential to damage cells in your body. So, for starters, increase the distance between your phone and your body. Don't tuck it into your pocket or wear it on a belt. Use a landline if available.

EMF exposure deceases rapidly as the distance from a cell phone increases. Don't bring your cell phone into your bedroom and put it on the night table. Shut it down or put it in another room. If this is not possible, put it on the other side of the room from your bed. Some distance is better than none.

Use your speakerphone for long calls, use earbuds, or even a Bluetooth headset or earpiece. But remove your Bluetooth headset as soon as you have finished your call. Text instead of calling whenever you can. The farther your phone is from your body, the better.

Don't use your cell phone when the signal is weak. If there are only one or two bars showing on the screen, the phone is emitting higher levels of radio frequencies to compensate. Wait until the signal improves to make your call. Also, don't make calls when you are in a moving vehicle, bus or train. Cell phones draw more power, and emit more radiation, in enclosed metal spaces.

Switch the phone to airplane mode when you are not using it, and especially if you give your cell phone to your kids to play with, especially toddlers. If possible, have your wireless router in a part of the house that is little-used, and out of the bedroom. Keep your bedroom as free of electronic radiation as possible.

Rethink your "smart" home devices. The more devices that are connected to your Wi-Fi or cellular network, the more EMF exposure you are going to have. This can include the smart home devices such as starting the oven and locking the door. Beware of radio-frequency-based smart meters which are being installed by utilities to control power consumption within a house.

Studies of Adverse Effects

"Waiting for high levels of scientific and clinical proof before taking action to prevent well known risks can lead to very high health and economic costs, as was the case with asbestos, leaded petrol and tobacco." —The European Commission

Even though the wireless industry and "experts" assure us electromagnetic fields (EMFs) from cell phones, other electronic devices like laptops, smart meters, Wi-Fi, and cell towers that generate non-ionizing radiation are not harmful, there is growing research which indicates EMFs may impact human health in negative ways. This section includes a selection of studies to demonstrate the variety of impacts that are being experienced.

The industry has pointed to studies which say there is no threat in the use of cell phones, even though the same studies point to the association of brain cancer in the case of "heavy use." (INTERPHONE 2010; Roosli M 2013; Coureau 2014; Carlberg 2015).   And in the case of 5G, it is impossible at this point to say it is not harmful because it is being developed as it is being implemented and there are few, if any, long term studies.

In critiquing long-term rat studies which have found a positive association with cancer, the cell industry points out the rats were exposed to "whole body" radiation rather than the limited radiation one might experience from holding a cell phone to one's ear. Yet we carry our cell phones in our pockets and access them for hours at a time for social media, texting, the Internet, and video games. And the "small cells" which will be mounted every 1000 ft on city streets will be broadcasting their electromagnetic fields constantly.

The Federal Communications Commission (FDA) and other agencies don't consider low-frequency EMFs harmful. However, the World Health Organization's (WHO) International Agency for Research on Cancer classified low-frequency EMFs as a "possible human carcinogen" based on human observational data and animal research that link EMF radiation to tumors (WHO IARC 2011).

Emerging research suggests we could sustain harmful effects of radiation at these frequencies, although the effects can take months or years to develop and depend on intensity, duration, frequency, and other variables.

The cell phone industry depends on the findings that non-ionizing radiation used in cell phones, etc., use does not cause appreciable heat response, and therefore is safe. But Pall (2015), and others show another mechanism by which low-frequency radiation causes harm. A growing body of research demonstrates that low frequency radiation disrupts "voltage-gated calcium channels" (VGCCs), transmembrane proteins that are found in many of our body's cells (Pall 2013; Pall 2015).

Catterall (2011) reports that in response to a potential cell membrane change, VGCCs—which act like gatekeepers for a cell—allow an influx of calcium ions into the cell to carry out core biologic processes, including muscle contractions, the release of hormones and neurotransmitters, gene expression, enzyme activity, and more.

More than 26 studies over the past 25 years have shown that non-ionizing radiation initiates a massive influx of calcium into cells and that this calcium influx can be blocked by calcium channel blockers, meaning that the mechanism is VGCC-dependent (Pall 2013, 2015).

Neurons, found in the brain as well as the body, have the highest density of VGCC of all human cell types, which means the brain and nervous system may be extra susceptible to EMF effects (Pall 2015). In animal studies, EMF exposure induced damaged neurons, brain structure changes, cognitive impairment, and brain inflammation (Kivrok 2017; Megha 2012; Salford 2003).

Excessive calcium influx into a cell can lead to oxidative stress (Pryor 1995; Lymar 2003), cellular DNA damage (Lai 2004; Szabo 2005; Phillips 2009), and apoptosis or cell death (Pall 2015; Kivrak 2017). All of these can affect the development of disease and cancer.

Di Ciaula (2018) did a research review detailing research findings that millimeter waves can alter gene expression, promote cellular proliferation and synthesis of proteins linked with oxidative stress, inflammatory and metabolic processes. The researchers conclude "available findings seem sufficient to demonstrate the existence of biomedical effects, and to invoke the precautionary principle."

Neufeld and Kuster (2018) documented how significant tissue heating can be generated by 5G technology's rapid short bursts of energy. "The results also show that the peak-to-average ratio of 1,000 tolerated by the International Council on Non-ionizing Radiation Protection (ICNRP) guidelines may lead to permanent tissue damage after even short exposures, highlighting the importance of revisiting existing exposure guidelines."

"The Human Skin as a Sub-THz Receiver—Does 5G Pose a Danger to it or Not? (Betzalel 2017) and "The Modeling of the Absorbance of Sub-THz Radiation by Human Skin (Betzalel 2018) present research that found higher 5G frequencies are intensely absorbed into human sweat ducts (in skin) at much higher absorption levels than other parts of our skin's tissues. The researchers conclude "we are raising a warning flag against the unrestricted use of sub-THz technologies for communication, before the possible consequences for public health are explored."

Bandara and Carpenter (2018) document the significant increase in environmental levels of radiofrequency (RF) electromagnetic wireless radiation over the last two decades. The study cites an evaluation that found 68.2% of 2266 studies in humans, animals, and plants demonstrated significant biological or health effects associated with exposure to electromagnet fields. 89% of experimental studies that investigated oxidative stress endpoints showed significant effects and that radiofrequency electromagnetic radiation causes DNA damage apparently through oxidative stress. Oxidative stress occurs if the equilibrium between formation of reactive oxygen species (ROS) and the capacity of antioxidant system to neutralize them is disturbed.

The study also highlights research that has associated RF exposure with altered neurodevelopment and behavioral disorders, structural and functional changes in the brain, and the sensitivity of pollinators. "These findings deserve urgent attention. This weight of scientific evidence refutes the prominent claim that the deployment of wireless technologies poses no health risks at the currently permitted non-thermal radio frequency exposure levels."

The European Cancer Environment Research Institute in Brussels, Belgium, reviewed current research findings in "Thermal and non-thermal health effects of low intensity non-ionizing radiation: An international perspective" (Belpomme 2018) and state "the mechanism(s) responsible include induction of reactive oxygen species, gene expression alteration and DNA damage through both epigenetic and genetic processes. They found that "exposure to low frequency and radio frequency electromagnet fields at low intensities pose a significant health hazard that has not been adequately addressed by national and international organizations such as the World Health Organization."

Oxidative Stress

What is oxidative stress caused by radio frequency electromagnetic fields, RF-EMFs? It was previously believed that as RF-EMFs do not have enough energy to remove electrons (producing an ion pair), they are unable to cause cancer. However, considering the evidence of free-radical damage which has been confirmed in studies on humans, animals, plants and microorganisms, some scientists now believe that EMF-induced oxidative stress can cause damage to cellular targets such as DNA and trigger processes which lead to cancer (Havas M 2017).

Kivrak (2017) explains despite being essential for life, oxygen molecules can lead to the generation of hazardous by-products, known as reactive oxygen species (ROS) during biological reactions. These reactive oxygen species can damage cellular components such as proteins, lipids and DNA. Antioxidant defense systems exist in order to keep free radical formation under control and to prevent their harmful effects on the biological system.

Oxidative stress occurs if the antioxidant defense system is unable to prevent the harmful effects of free radicals. In a review of studies highlighting the impact of oxidative stress on antioxidant systems, Kivrak reports several studies found that exposure to EMF results in oxidative stress in many tissues of the body. Exposure to EMF is known to increase free radical concentrations (Kivrak 2017).

Considering the evidence of free-radical damage which has been confirmed in studies on humans, animals, plants and microorganisms, some scientists now believe that EMF-induced oxidative stress can cause damage to cellular targets such as DNA and trigger processes which lead to cancer (Havas 2016).

Nielson (2019) states, "Weak radio frequency (RF) magnetic fields in the MHz-range was shown to influence the concentrations of reactive oxygen species (ROS) in living cells (Usselman 2016; Usselman 2014; Castello 2014; Naarala 2017). Oxidative stress denotes high levels of ROS that incur damage to DNA, protein or lipids.

Cellular oxidation is a well-studied potential mechanism of harm from radiofrequency radiation. Healthy biological systems require a balance of oxidation and anti-oxidation to fight infection and prevent disease. A review of the literature by Yakymenko (2016) confirmed that in 93 of 100 studies, non-ionizing radio-frequency radiation caused a cellular stress response with excessive reactive oxygen species (ROS). He concluded, "oxidative stress induced by RFR exposure should be recognized as one of the primary mechanisms of the biological activity of this kind of radiation."

Microwave and millimeter wavelength radiation, which is one of the key blocks of spectrum that 5G service will use, is also considered non-ionizing and is purported to not produce the kind of energy that directly damages cells, according to the cellular industry. Common devices, such as Wi-Fi routers, garage door openers, airport security scanners and walkie-talkies, use lower frequency microwaves.

But some experts suspect that the radiation from these devices could damage cells via another biological mechanism, such as oxidative stress in cells, which leads to inflammation and has been found to cause cancer, diabetes and cardiovascular, neurological and pulmonary diseases.

Reactive oxygen species (ROS) are a normal part of cellular processes and cell signaling. Overproduction of

ROS that is not balanced with either endogenous antioxidants (superoxide dismutase, catalase, glutathione peroxide, glutathione, melatonin) or exogenous antioxidants (Vitamin C, Vitamin E, carotenoids, polyphenols) allows the formation of free radicals that oxidize and damage DNA, proteins, membrane lipids and mitochondria (Russell 2018a).

Mitochondrial DNA is more susceptible to DNA damage than nuclear DNA as it lacks histones, has a reduced ability to repair DNA, and is not protected from mitochondrial reactive oxygen species (Gorlach 2015).

Oxidative damage from ROS has been increasingly linked to the development and/or exacerbation of a number of chronic diseases and cancer (Thannickal and Fanburg 2000; Valko 2006; Alfadda and Sallam 2012).

Erdal (2008) conducted a test on rats, "Effects of long-term exposure of extremely low frequency magnetic field on oxidative/nitrosative stress in rat liver." The conclusion suggested the long-term extremely low frequency magnetic field (ELF-MF)

exposure may enhance the oxidative/nitrosative stress in liver tissue of the female rats and could have a derivative effect on cellular proteins.

Santini (2018) reviewed the existing literature regarding the effects of EMFs on reproductive systems. Given the role of mitochondria as the main source of reactive oxygen species (ROS), they focused on the hypothesis of a mitochondrial basis of EMF-induced reproductive toxicity. Numerous studies revealed the detrimental effects of EMFs from mobile phones, laptops, and other electric devices on sperm quality and provide evidence for extensive electron leakage from the mitochondrial electron transport chain as the main cause of EMF damage. In female reproductive systems, the contribution of oxidative stress to EMF-induced damages and the evidence of mitochondrial origin of ROS overproduction are reported as well.

"900-MHz microwave radiation promotes oxidation in rat brain" (Kesari 2011) exposed Wistar rats to a mobile phone for 2 hours per day for a duration of 45 days where the specific absorption rate was 0.9 W/Kg. The study concluded that a reduction or an increase in anti-oxidative enzyme activities, protein kinase C, melatonin, caspar 3, and creatine kinase are related to overproduction of reactive oxygen species (ROS) in animals under mobile phone radiation exposure, and that the findings on these biomarkers are clear indications of possible health implications.

Reproductive Health

The Centers for Disease Control and Prevention (CDC) report almost 1 in 8 couples will experience some degree of infertility in their lifetime (CDC 2019). Men are often told to keep laptops off their lap to prevent overheating and damaging a sensitive area, but it appears that EMFs may induce non-thermal, harmful effects on male fertility.

A study on male infertility, "Radiations and male fertility," (Kesari 2018) focused on radiation deriving from cell phones, laptops, Wi-Fi and microwave ovens. The study concluded that RF-EMF may induce oxidative stress with an increased level of reactive oxygen species (ROS), which may lead to infertility. This was concluded based on available evidences from *in vitro* and *in vivo* studies suggesting that RF-EMF exposure negatively affects sperm quality.

In "Effect of 2.45 GHz microwave radiation on the fertility pattern in male mice," the aim of the study was to explore the effects of microwave radiation exposures on 6-8 weeks old male Swiss albino mice (Jonwal 2018). It concluded that 2.45 GHz microwave radiation exposure causes oxidative stress in testes and it may lead to detrimental and injurious effects on the fertility potential of the male reproductive system of Swiss albino mice.

Sprague-Dawley male rats were used to investigate the effect of long duration exposure to electromagnetic field from mobile phones on spermatogenesis in rats using 4G-LTE levels (Oh 2018). The researchers found that the longer exposure duration of electromagnetic field decreased the spermatogenesis, calling for further investigations on the potential effects of EMF from mobile phones on male fertility.

"Effect of radio frequency radiation on reproductive health" (Singh 2018), documents research that has found a link between radio frequency radiation and oxidative stress and changes to the reproductive system including sperm count, motility, normal morphology and viability. The review concludes the "available data indicate that exposure to EMF can cause adverse health effects. It is also reported that biological effects may occur at very low levels of exposure."

Exposure to 900 MHz EMF causes alterations in adult rat testicular morphology and biochemistry was shown in a study entitled, "Exposure to a 900 MHz electromagnetic field for 1 hour a day over 30 days does change the histopathology and biochemistry of the rat testis" (Odaci 2015).

Kumar (2014) reports a significant decrease in sperm count, increase in the lipid peroxidation damage in sperm cells, reduction in seminiferous tubules and testicular weight, and DNA damage were observed following exposure to EMF in male albino rats. The author concludes that mobile phone exposure adversely affects male fertility.

In a classic 1997 study, "RF radiation-induced changes in the prenatal development of mice," (Magras and Xenos 1997), twelve pairs of mice, divided in two groups, were placed in locations of different power densities and were repeatedly mated five times. There was a progressive decrease in the number of newborns per generation, and the mating ended in irreversible infertility at the fifth generation.

After exposure to cell phone radiation, human semen had higher levels of oxidation, as well as decreased sperm motility and viability (Agarwal 2009; Erogul 2006). EMF exposure has also been correlated with sperm DNA damage, lower testosterone, erectile dysfunction, and abnormal sperm structure in rodents and humans (Kumar 2013; Kumar 2014; Al-Ali 2013).

"Effect of radio frequency radiation on reproductive health," (Singh 2018) documents research that has found a link between radio frequency radiation and oxidative stress and changes to the reproductive system including sperm count, motility, normal morphology and viability. The review concludes that the "available data indicate that exposure to EMF can cause adverse health effects. It is also reported that biological effects may occur at very low levels of exposure."

The effect of long-term radiation of mobile phones (1800 MHz) on female reproduction and oxidative stress in mice was investigated by Shahin (2017). The data of the study indicated that mobile phone radiation is able to increase the levels of reactive oxygen species (ROS), along with a variety of other results. The data attested to the harmful impact of mobile phone's radiation on female fertility.

In "Extremely low-frequency electromagnetic field (EMF) generates alterations in the synthesis and secretion of estradiol-17B (E2) in uterine tissues: An *in vitro* study," (Koziorowska 2018) the researchers' aim was to determine the effect of an EMF on the synthesis and secretion of estrogen in the (porcine) uterus. They concluded the EMF induces changes in the synthesis and release of estrogen in uterine tissues during the estrous cycle.

Children

Children may be extra susceptible to EMFs. Babies, small children and young people have thinner skin and bones, and higher water content in their tissues, which might concentrate the effect of EMFs in their bodies compared to fully grown adults (Morgan 2014). A few earlier studies have found disturbing connections between childhood cancer and close proximity to radio stations, high-voltage power lines and high-EMF environments (Michelozzi 2002; Svendsen 2007).

More recently, epigenetic studies have found some neurodevelopmental and neurobehavioral changes due to exposure to wireless technologies. Symptoms of retarded memory, learning, cognition, attention, and behavioral problems have been reported in numerous studies and are similarly manifested in autism and attention deficit hyperactivity disorders, as a result of EMF and RFR exposures where both epigenetic drivers and genetic (DNA) damage are likely contributors (Sage 2018).

Epigenics is the way in which the expression of heritable traits is modified by environmental influences or other mechanisms without a change to the DNA sequence. The authors suggest technology benefits can be realized by adopting wired devices, rather than wireless, for education to avoid health risk and promote academic achievement.

The study, "Absorption of wireless radiation in the child versus adult brain and eye from cell phone conversation or virtual reality" (Fernandez 2018), points out that children's brains are more susceptible to hazardous exposures, and are thought to absorb higher doses of radiation from cell phones in some regions of the brain. Unfortunately, since 1997, safety testing has relied only on a large, homogenous, adult male head to simulate exposures.

The authors report that modeling of a cell phone held to the ear, or of virtual reality devices in front of the eyes, reveals that young eyes and brains absorb substantially higher local radiation doses than adults. They call for the need to apply refined methods for regulatory compliance testing; for public education regarding manufacturers' advice to keep phones off the body, and prudent use to limit exposures, particularly to protect the young.

A study of kindergarten children in Australia, "Radiofrequency-electromagnetic field exposures in kindergarten children" (Bhatt 2017), showed that environmental RF-EMF exposure levels exceeded the personal RF-EMF exposure levels at kindergartens. They computed statistics of RF-EMF exposures for 16 frequency bands between 88 MHz and 5.8 GHz.

In the 2016 paper, "Controversies on electromagnet field exposure and the nervous systems of children" (Warille 2016), researchers reviewed possible health effects from exposure to low levels of electromagnetic field (EMF) in children, arising from electrical power sources and mobile phones. When the current data was considered in detail, it was noted that children's unique vulnerabilities make them more sensitive to EMFs emitted by electronics and wireless devices, as compared to adults.

Warille found some experimental research shows a neurological impact, and exposure in humans may lead to cognitive and behavioral impairments. The researchers called for new research approaches which focus on the effects on the developmental processes of children exposed to electromagnetic fields, using consistent protocols.

Radiofrequency electromagnetic field (RF-EMF) exposure regulations and guidelines generally only consider acute effects, and not chronic, low exposures. Redmayne (2016) writes that concerns for children's exposure are warranted due to the amazingly rapid uptake of many wireless devices by increasingly younger children in "International policy and advisory response regarding children's exposure to radio frequency electromagnetic fields (RF-EMF)" (Redmayne 2016).

Research provides a representative cross-section of policy and advisory responses within set boundaries. There are a wide variety of approaches ranging from ICNIRP/IEEE (International Commission on Non-ionizing Radiation Protection/Institute of Electrical and Electronics Engineers) guidelines and "no extra precautions needed" to precautionary or scientific much lower maximums and extensive advice to minimize RF-EMF exposure, to banning advertising/sale to children, and proposals to add exposure information to packaging. The wide range of policy approaches can be confusing for parents and those who take care of children.

Redmayne calls for some consensus among advisory organizations, acknowledging that, despite extensive research, the highly complex nature of both RF-EMF and the human body, and frequent technological updates, means the simple assurance of long-term safety cannot be guaranteed. The author recommends minimum exposure of children to RF-EMF, and that the ICNIRP guidelines need to urgently publish how the head, torso, and limbs' exposure limits were calculated, and what safety margin was applied, since this exposure, especially to the abdomen, is now dominant in many children.

Pooled analysis demonstrated a two-fold risk increase of childhood leukemia as a result of exposures to extremely low frequency EMF, along with the consistency of studies across different countries, types of study design, methods of exposure assessment, and systems of power transmission and distribution (Schuz 2011).

DNA damage

Oxidative stress occurs if the equilibrium between formation of reactive oxygen species (ROS) and the capacity of antioxidant system to neutralize them is disturbed. The uncontrolled overproduction of ROS can lead to DNA damage such as single/double-strand breaks and crosslinks (Wells 2015).

Over the past few decades, it has been revealed that EMFs at extremely low frequencies are capable of increasing the production of free radicals, which can cause DNA double-strand breaks (Anderson 1993; Buldak 2012; Consales 2012; Du 2008; Esmaeili 2017; Giorgi 2011; Jouni 2012; Tkalec 2007; Yokus 2005).

Lai and Singh (1997) reported that exposure of rats' brain cells to a 60 Hz magnetic field induced DNA single and double bond breaks. In another work, Mihai (2014) demonstrated that extremely low-frequency EMF are capable of inducing DNA strand breaks in normal human cells. Mihai hypothesized that the underlying mechanism of detected DNA damage induction is the production of ROS induced by EMF.

Duan (2015) studied a potential genotoxicity of 50 Hz extremely low-frequency EMF and 1800 MHz radiofrequency EMF on mouse spermatocyte-derived GC-2 cell line. The results showed a significant increase in DNA strand breaks in the cells exposed to extremely low-frequency EMF. 1800 MHz radiofrequency is associated with mobile telecommunications and is widely used worldwide.

Franzelletti (2010) scrutinized the ability of high frequency EMF to cause DNA breaks. The authors irradiated trophoblast cells with a 1.8 GHz continuous wave EMF as well as various GSM signals, which are mobile phone standard. The high frequency EMF caused transient increases of DNA fragmentation (separation of DNA strands into pieces).

Panagopulus (2007) detected DNA fragmentation induced by EMF emitted from mobile phones. The researchers utilized Drosophila melanogaster (common fruit fly) as a biological model for studying the effect of EMF from common mobile phone (GSM 900 MHz and DCS, 1800 MHz standards) signals on Drosophila oogenesis (ovum). It was demonstrated that the radiation from both types of mobile phones signals induced DNA fragmentation and cell death.

It was revealed that EMF can initiate the displacement of electrons in DNA, which is accompanied by electron transfer, in a study by Blank and Goodman (2008). The displacement directly affects the hydrogen bonds, which are responsible for DNA integrity and spatial DNA geometry. As a result, the DNA chain undergoes separation and transcription.

There is an abundance of publications indicating a potential genotoxic effect of EMF operating at radiofrequency band, in particular, an impact of radiation of mobile phones on human DNA. Mazor (2008) scrutinized the effects of exposure of human lymphocyte to radiofrequency EMFs (800 MHz, continuous wave) on genomic instability. The specific absorption rates (SARs) were 2.9 and 4.1 W/kg—that is close to the current levels established by the ICNIRP guidelines for mobile phones.

The exposure resulted in an increase of the levels of aneuploidy. Aneuploidy is a signature of chromosal abnormality, which has been strongly associated with miscarriage and birth defects in humans. At a cellular level, aneuploidy has been linked to cell malfunction and carcinogenesis.

In another work dedicated to studies of the impact of mobile phones on DNA, human peripheral blood lymphocytes were exposed to continuous 830 MHz EMF, which is in the range of GSM standard telephony (Mashevich 2003). A linear increase in chromosome 17 aneuploidy as a function of the SAR value indicating a genotoxic effect of radiofrequency EMF was detected.

A 2019 study on human blood found microwave 3G EMF/radiation--within the current exposure limits--had significant genotoxic action on human cells (Panagopoulos 2019). A series of landmark papers found that effects from microwaves on human lymphocytes can be dependent on carrier frequency (Belyaev 2005), that UMTS (3G) microwaves can affect chromatin and inhibit formation of DNA double-strand breaks (Belysev 2009), and that stem cells are most sensitive to microwave exposure. Children have more active stem cells than adults.

Cancer

Several studies have investigated whether mobile phone use is associated with an increased risk for gliomas (malignant tumors) (Lahkola 2008; Klaeboe 2007; Christensen 2005; Lonn 2005; Coureau 2013; Hardell 2013b; Cardis 2011). The results were not conclusive. However, a recent study analysis of two case-control studies on malignant brain tumors, that included 1498 cases and 3530 controls, revealed that mobile phone use increased the risk of glioma. The risk was nearly double in the group with over 25 years latency period. An increased risk was also found for cordless phone use. Furthermore, in the same study it was of note that the temporal lobe had the highest risk for glioma occurrence (Hardell 2013b).

A study conducted in France that included 253 gliomas, 194 meningiomas and 892 matched controls, revealed no association between mobile phone use and risk for gliomas. Illogically, however, the researchers found that in heavy mobile phone users with cumulative duration, there was a significant increased risk for glioma (Coureau 2014).

In 2010, Hardell reported an increased risk for glioma for both short and long term mobile phone users. The report included patients 20-80 years old which is important because the highest incidence of glioblastoma, the most common and malignant brain tumor, is found in the age group 45-75 years (Hardell 2010).

The large INTERPHONE study was conducted in 13 countries with 16 centers. The studies included patients 30-59 years. The study concluded an absence of increased risk of glioma with use of mobile phones. Again, there was an increased risk of glioma at the exposure level of ten years or more. Rather than make an association of risk for heavy use, they simply stated further investigation was needed in order to draw safe conclusions (INTERPHONE 2010).

A more recent pooled analysis of two Swedish case-control studies on 1625 meningioma patients and 3530 control patients showed a relative increased risk among heavy users of mobile and cordless phones (Carlberg 2015). Of concern is the fact that meningioma is a slow-growing tumor and a longer latency period is needed in order to draw definitive conclusions.

Acoustic neuromas are also slow-growing tumors, therefore the observation period should not be short. For acoustic neuroma, a large prospective study reported an increased risk with long term use compared to patients who never used a mobile phone. Furthermore, the risk increased with increased duration of use (Benson 2013).

There are many studies that do show a causal relationship between EMFs and cancer. Because of the brain's high metabolic rate, it is more prone to oxidative damage (which EMFs may cause) compared to other organs (Ozmen 2007). Several papers published as part of the Hardell group of studies in Sweden found an increased risk for brain cancer in people who used cell phones, particularly on the same side of the head that the phone was used on (Hardell 2002, 2006, 2009, 2011).

Miller (2018) gave a cancer epidemiology update with a comprehensive research review of RF effects in human and animal research. The review concludes that scientific evidence is now adequate to conclude radio frequency radiation is carcinogenic to humans. Previously published studies also concluded that RF can "cause" cancer (Carlberg and Hardell 2017; Atzmon 20106; Peleg 2018).

The US National Toxicology Program (NTP) Study on Cell Phone Radiation found "clear evidence" of cancer and DNA damage in a $25-million study designed to test the basis for federal safety limits (NTP 2018). The heart and brain cancers found in the NTP rats are the same cell type as tumors that researchers have found to be increased in humans who have used cell phones for over ten years. The researchers say this animal evidence confirms the human evidence associating the exposure to cancer (Hardell 2019).

"The Ramazzini Institute Study on Base Station RF" was another large-scale rat study that also found increases in the same heart cancers that the NTP study found—yet the Ramazzini rats were exposed to much lower levels of RF than the NTP rats. In fact, all the Ramazzini radiation exposures were below FCC limits, as the study was specifically designed to test the safety of RF limits for cell tower/base stations (Falconi 2018). The Ramazzini study corroborates the NTP findings.

"Tumor promotion by exposure to radiofrequency electromagnetic fields below exposure limits for humans" is a replication study that used very, very low RF exposures (lower than the Ramazzini and NTP study). Researchers found elevated lymphoma and significantly high numbers of tumors in the lungs and livers in the animals (Tilman 2010), leading researchers to state that previous research was confirmed and that "our results show that electromagnetic fields obviously enhance the growth of tumors" (Lerchi 2015).

Prasad (2017) investigated the results of 22 case-controlled studies which showed an increased risk of brain tumor with long-term exposure to mobile phone radiation, in which industry-funded research tended to underestimate the risk.

In the first court case to award damages to a plaintiff for a brain tumor caused by cell phones, an Italian court excluded cancer-based studies related to cellphones that had been financed by telecommunications companies, according to a news article (Williams, 2017).

Eyes and vision

There is particular concern for 5G applications as the eyes would also receive significant radiation especially for near field exposures. Cataracts remain the leading cause of blindness in the world and are a societal burden due to the high incidence, cost and consequences to quality of life (CDC 2015).

From the available literature, it appears that microwave frequencies including MMW proposed for 5G can have non-thermal biological effects on the lens of the eye. 5G deployment will add shorter wavelengths to longer wavelengths which have not been adequately tested for long term exposure. With the expected rise of wearable ocular (pertaining to the eyes) digital technology devices such as virtual reality for gaming, entertainment, the social sciences and healthcare, there will be significantly more exposure to microwave radiation very close to the orbit.

National Institutes of Health (NIH) statistics from 2010 show there is a 17.11% overall prevalence of cataracts over age 40 (NIH NEH 2010) and a steady rise in cataract surgeries (Goilogly 2013).

Well established risk factors in the development of cataracts are age, smoking, diabetes, and UVB exposure. Now research is pointing towards oxidative damage as a general mechanism for age related cataracts (Spector 1995; Ye 2001; Abraham 2007). Microwave radiation is also a known cause of cataracts with that being an undisputed mechanism. The eyes lack sufficient blood flow to dissipate heat effectively. There is some evidence that repeated low level exposures to microwave radiation could cause cataracts, but researchers agree that more studies are needed (Vignal 2009; Carpenter and Van Ummeren 1968; Moss 1977; Foster 1986; Van Umersen and Cogan 1976; Riva 2005; Ryzhov 1991; Drean 2013; Morgan 2015).

In a 2014 publication in the Institute of Electrical and Electronics Engineers journal, IEEE Transactions on Microwave Theory and Techniques, Sasaki (2013) reported on their *in vivo* rabbit experiments for operating frequencies ranging from 24.5 to 95 GHz, measuring temperature elevation. Their studies suggest that corneal damage occurred at an incident power density of 300 mW/cm2. They conclude that ocular heating should be the basis for safety guidelines for near field exposure. It is mentioned, however, that only a few experimental studies in the millimeter wavelengths were used to determine the current exposure guideline limits.

In another IEEE publication looking at MMW health effects, Wu (2015b) supports current standards of safety based on heat, but also point out that the MMW research on biological effects is sparse relative to that of longer microwave frequencies. They advise that additional studies may be needed to examine the potential biological effects of MMW radiation in order to develop appropriate consumer guidelines, especially where antennas are located close to the body.

Yu and Yao (2010) reviewed literature on microwave radiation and induction of cataracts. Reports of non-thermal biologic effects of microwave radiation include alteration of cell proliferation and apoptosis, inhibition of gap junctional intercellular communication, stress response and genetic instability.

Bormusov (2008) examined the non-thermal effects of high frequency radiation from cell phones and other wireless devices on lens epithelium. They found both reversible and irreversible ocular changes and note that the effects they saw with short term exposure at low levels could translate to similar effects with cataracts over a 10—20-year period of cumulative exposure. They state, "It is recommended to use cell phones from a distance to minimize exposure, thus reducing any potential harmful effects of cell phone use on the lens."

Prost (1994) was one of the first to study the effects of millimeter microwave radiation on the eye. He noted that microwaves of different wavelengths have been implicated in the development of cataracts. His research found that low power millimeter waves produced lens opacity in rats over a 58-day period indicating MMW is a predisposing factor for cataracts.

Kues and Monhan (1992) at John Hopkins University researched the effects of low-level microwave radiation on the primate eye using 1.25 and 2.45 GHz wavelengths for 4 hours daily for 3 consecutive days. They identified damaging ocular effects including corneal lesions, increased vascular permeability and degeneration of photoreceptors in the retina. They found that pulsed microwave exposure produced abnormalities at lower power densities than continuous wave exposure. These were relatively short exposure periods.

Cutz (1989) in his publication "Effects of microwave radiation on the eye: The occupational health perspective," looked at occupational exposure to RF EMR, noting that eye effects from microwave radiation can be thermal or non-thermal and that lens opacities can be generated experimentally in animals with relatively high intensity RF EMR. He states that for lower intensities cumulative exposures may cause damage. He also reported that microwaves caused degeneration of retinal nerve endings. Long term effects were not determined, pointing to the need for additional research.

Lipman (1988) noted that microwaves most commonly cause anterior and/or posterior subcapular lenticular opacities both in experimental animals, epidemiological studies and case reports. They indicate that cataract formation is related to the power of the microwave radiation and duration of exposure. Lipman concludes that until further definitive research is conducted on the mechanisms of injury and protective measures identified, mechanical shielding is recommended to minimize the possibility of development of radiation-induced cataracts.

Current safety guidelines are based on heat measurements. The absence of current literature on the effects on the eye of millimeter wavelengths highlights the need for much more independent research and precaution moving forward to prevent an epidemic of pathology.

Heart

Some people who are experiencing electrosensitivity complain that they have a rapid or irregular heartbeat, and feel chest pressure or pain as a part of their overall experience (Eltiti 2007).

In an early study, Glotova and Sadchikova (1970) published "Development and clinical course of cardiovascular changes after chronic exposure to microwave irradiation; Effect of microwave exposure to microwave irradiation." The purpose of the report was to describe the nature, severity, and clinical course of the cardiovascular changes that follow chronic exposure to microwave irradiation. This data was determined from 3 to 6 years of clinical observations on 130 patients.

The patients had been working with microwaves in the one-centimeter range waves for at least 5 years and were exposed to fairly intense levels, especially in the early years.

Two groups were studied within this study. Group one had asthenia (weakness and low energy), and complained of headache, fatigue, insomnia and pains in the heart region. A number of these persons had arterial hypotension (low blood pressure) and bradycardia (slow heart rate).

Group two complained of fatigue, irritability, headaches, nausea, and vertigo. Some experienced constricting pain in the heart and subjects in this group were more likely to experience tachycardia (a rapid heartbeat).

The authors concluded that while some individuals exhibit only mild asthenic systems with sinus bradycardia, others develop autonomic-vascular dysfunction, often with symptoms of hypothalamic insufficiency and angiospasm (the spasmodic contraction of the blood vessels with increase in blood pressure) which sometimes impair the cerebral (brain) and coronal (heart) circulation.

A NASA document, "Electromagnetic field interactions with the human body; Observed effects and theories" (NASA 1981) highlighted electrosensitivity (also known as electrohypersensitivity, EHS or microwave illness) symptoms. Electrosensitivity is an acute symptom complex in some 3% of individuals who experience heart palpitations, headaches, nausea, irritability in the presence of a cell phone, Wi-Fi router, cell tower or other wireless devices.

Chernyakov (1989) induced heart rate changes in anesthetized frogs by microwave irradiation of remote skin areas. Complete denervation of the heart did not prevent the reaction. This suggested a reflex mechanism of the MMW action involving certain peripheral receptors.

Potckhina (1992) found certain frequencies from 53 to 78GHz band continuous wave changed the natural heart rate variability in anesthetized rats. He showed that some frequencies had no effect (61 or 75 GH) while other frequencies (55 and 73 GHz) caused pronounced arrhythmia. There was no change in skin or whole-body temperature.

Pakhomov (1998) looked at the biological effects of MMWs. He examined dozens of studies and cites research demonstrating profound effects of MMW on all biological systems including cells, bacteria, yeast, animals and humans. Some effects were clearly thermal, however, many of the studies showed non-thermal biological effects at low intensities. He included there were effects on heart rate variability. Both negative and positive responses were seen depending on frequency, power, resonance and exposure time.

Havas (2010) conducted a "proof of concept" study to determine if heart rate changes caused by microwave radiation could be measured with real-time monitoring. They found that some individuals developed a rapid or an irregular heartbeat when exposed to pulsed microwaves from a cordless phone base station at levels considered safe by the WHO, FCC, and Health Canada. It was part of a collection of scientific research papers published in the European Journal of Oncology, as a special International Commission for Electromagnetic Safety (ICEMS) Monograph entitled "Non-thermal Effects and Mechanisms between Electromagnetic Fields and Living Matter" (Giuliani 2010).

In another study, "Radiation from wireless technology affects the blood, the heart, and the autonomic nervous system," Havas (2013) writes about the symptoms of electrohypersensitivy (EHS), best described as rapid aging syndrome, and includes heart palpitations, pain or pressure in the chest accompanied by anxiety among the many symptoms. The article demonstrated that the response to "electrosmog," (the electromagnet radiation emitted by all the computers, mobile phones, and all other artificially generated electromagnetic fields in the environment) is physiologic and not psychosomatic.

Elmas (2016) presented a systematic review of collective scholarly literature examining the effects of EMFs on the heart. It's noted that although most works describing effects of EMF exposure have been carried out using electric frequencies of 50-60 Hz, a consensus has not been reached about whether long- or short-term exposure to those frequencies negatively affects the heart. At greater

EMF strengths or shorter exposures, the ability of the body to develop compensation mechanisms is reduced and the potential for heart-related effects increases.

There is the dichotomy that diseases of heart tissues such as myocardial ischemia can also be successfully treated using EMF, but the study concludes "until the effects of EMF on heart tissue are more fully explored, electronic devices generating EMFs should be approached with caution."

Bandara and Weller (2017) point out the strong and growing evidence that radio frequency radiation from wireless devices can negatively affect normal functioning of cellular processes throughout the body via oxidation/inflammation pathways. They started with military studies that showed cardiovascular effect and ended with a review of 242 studies on RFR and oxidative stress noting, "a staggering 216 (89%) of them found significant effects".

The Bandara and Weller (2017) study begins with a reference to Vernon (2017), "Cardiovascular disease: Time to identify emerging environmental risk factors." Vernon reports a significant increase in the proportion of first ST elevation myocardial infarction (STEMI) patients without standard modifiable cardiovascular risk facts such as high cholesterol, high blood pressure, diabetes and smoking. They called the attention of researchers in cardiovascular disease (CVD) to emerging environmental risk factors, focusing on microwave radio frequency electromagnet radiation (RF-EMR).

They point out human exposure to RF-EMR has exponentially increased over the past three decades due to rapid and widespread deployment of wireless communication and surveillance infrastructure and the use of personal wireless devices, calling RF-EMR an environmental pollutant with cytotoxic effects.

2G radiation was found to cause DNA damage and an increase in the risk of cancer of the heart, brain and adrenal medulla in the recent 10-year, $25 million dollar National Toxicology Program study (NTP 2018). The report revealed not only clear evidence of cardiac tumors and DNA damage, but also histopathologic changes in the heart consistent with aging.

A current study, Adebayo (2019) aimed at determining the consequence of RF-EMR (1800 MHz) on the histological, hematological and histochemical properties of selected tissues of rat, along with assessing morphological changes associated with such exposures. As part of the results, the heart showed gross distortion of cardiac muscular architecture with distorted irregular cardiac muscle fibers and wider interfibres spaces.

Immune system

Our immune system is essential for our survival. Without an immune system, our bodies would be open to attack from bacteria, viruses, parasites, and more. It is our immune system that keeps us healthy as we drift through a sea of pathogens.

A 2019 study, "Effect of cell phone radiation on neutrophil of mice," (Yinhui 2019) evaluated the effect of cell phone radiation on neutrophil (white blood cells) of mice. No significant differences were seen in the short term, while the numbers of neutrophils were seen to rise after an exposure of 4 or 6 weeks. The authors concluded mobile phone radiation could give rise to an increase of neutrophil numbers, and radiation was likely to cause a decrease of phagocytosis and induced apoptosis (cell death) of neutrophils.

Lasalvia (2018) completed a study where they investigated the occurrence of biochemical/biological modifications in human peripheral blood lympho-monocytes exposed in a reverberation chamber. Times ranged from 1 to 20 hours. They found enlarged and deformed cells after EMF exposure and the onset of biochemical modification, mainly consisting in the reduction of the DNA backbone-linked vibrational modes.

The results suggest the occurrence of adaptive mechanisms put in action to compensate for early impairments of the oxidative phosphorylation system caused by exposure to EMFs. Oxidative phosphorylation is the system by which energy is obtained by electron transport and takes place in the mitochondria (the cell powerhouse) during aerobic respiration.

The authors conclude that overall the data presents a warning for health safety of people involved in long-term exposure to electromagnetic fields, although further studies are required to pinpoint the leukocyte cellular subset(s) selectively targeted by the EMFs action and the mechanisms by which it is achieved.

Zoothansiama (2017) evaluated the effect of radiofrequency radiations emitted in the vicinity of mobile phone base stations on the DNA damage and antioxidant status. The analyses of data from the group residing within a perimeter of 80 meters of mobile base stations showed significantly higher frequency of micronuclei when compared to the control group residing 300 meters away from the mobile base stations. Micronuclie arise from chromosome fragments as a result of chromosome breaks. These fragments fail to become incorporated into daughter cell nuclei during cell division.

El-Gohary (2017) studied the effect of electromagnetic waves from mobile phones on the immune status of male rats. They assessed the effect of electromagnetic field emitted from a mobile phone on the immune system in rats and the possible protective role of vitamin D. After 30 days of exposure time, 1 h/day, exposure resulted in a significant decrease in immunoglobulin. Immunoglobulin is any of a class of proteins present in the serum and cells of the immune system, which function as antibodies. They attach to foreign substances and assist in destroying them.

It also resulted in a significant increase in neutrophil and monocyte count. These changes were more increased in the group exposed to 2 hour/day. The authors concluded that exposure to mobile phone radiation compromises the immune system of rats, and vitamin D appears to have a protective effect.

The immune system is a biological indicator sensitive enough to measure the "weak biological influence" of RF/MWs. It is able to react in a measurable way to discrete environmental stimuli. In "Reaction of the immune system to low-level RF/MW exposures" (Szmigielski 2013), discussed the impacts of weak RF/MW fields, including cell phone radiation, on various immune functions, both *in vitro* and *in vivo*.

They found the bulk of available evidence clearly indicates that various shifts in the number and/or activity of immunocompetent cells are possible. They conclude "Certain premises exist which indicate that, in general, short-term exposure to weak MW radiation may temporarily stimulate certain humoral [relating to body fluids] or cellular immune functions, while prolonged irradiation inhibits the same functions."

Kolomytseva (2002) looked at the function of peripheral blood neutrophils under whole-body exposure of healthy mice to low-intensity extremely high-frequency electromagnetic radiation (EHF EMR 42.0 GHz, 0.15 mW/cm2, 20 min daily). The study showed 50% suppression of phagocytic activity of neutrophils after a single exposure to MMW radiation with the authors noting a profound effect on nonspecific immunity.

Lushnikov (2003) investigated cell-mediated immunity and nonspecific inflammatory response in mice exposed to low-intensity extremely high-frequency electromagnetic radiation. They found that MMW radiation reduced both immune and nonspecific inflammatory responses.

Gapeev (2003) showed for the first time that low-intensity extremely high-frequency MM electromagnetic radiation *in vivo* causes effects on spatial organization of chromatin in cells of lymphoid organs which is part of the

circulatory system and the immune system. Chromatin is
a complex of DNA and proteins that forms chromosomes
within the nucleus of eukaryotic cells. Most living things
are made up of eukaryotic cells, having a distinct nuclei and
chromosomes that contain their DNA. He exposed mice
to a single whole-body exposure for 20 min at 42.0 GHz
and 0.15 mW/cm2. He suggests that the effects were due
to involvement of the neuroendocrine and central nervous
systems.

Bacterial antibiotic resistance

Said-Salman (2019) presented a study which evaluated the non-thermal effects of a Wi-Fi router operating at 2.4 GHz on different pathogenic bacterial strains—*Escherichia coli, Staphylococcus aureus,* and *Staphylococcus epidermis.* They were exposed continuously for 24 and 48 hours. The exposure to Wi-Fi radiation altered motility and antibiotic susceptibility of *Escherichia coli,* however, there was no effect on the susceptibility of *Staphylococcus aureus,* and *Staphylococcus epidermis.* On the other hand, the exposed cells, as compared to the unexposed control cells, showed an increased metabolic activity and biofilm formation ability in *Escherichia coli, Staphylococcus aureus,* and *Staphylococcus epidermis.*

A published study entitled, "Effect of Mobile Tower Radiation on Microbial Diversity in Soil and Antibiotic Resistance" (Sharma and Lamba 2018) took soil samples from four different base stations located in Dausa City [India], and control samples from soil far from stations, and then isolated and evaluated the microorganisms in the soil. The researchers found greater antibiotic resistance in microbes present in soil near base stations compared to the control. The study concludes, "our findings suggest that mobile tower radiation can significantly alter the vital systems in microbes and turn them multi drug resistant (MDR) which is the most important current threat to public health."

Soghomonyan (2016) found that MMW affected growth and antibiotic sensitivity of *E. coli* and many other bacteria via non-thermal mechanisms. This may lead to antibiotic resistance.

Cocci are the leading pathogens of humans. It is estimated that they produce at least a third of all the bacterial infections of humans, including strep throat, pneumonia, otitis media (inflammation of the ear), meningitis, food poisoning, various skin diseases and severe types of septic shock. The effect of electromagnetic field (EMF) exposures at the microwave (MW) frequency of 18 GHz, on four cocci, *Planococcus maritimus*, *Staphylococcus aureus*, *S. aureus*, and *S. epidermidis* was investigated by Nguyen (2015). They demonstrated that exposing the bacteria to an EMF induced permeability in the bacterial membranes of all strains studied.

Torgomyan and Trchounian (2013) reviewed research on the mechanisms of bactericidal and antibiotic resistance after exposure to low intensity MMW. They suggest that alternations in water structure, cell membrane or the genome leading to changes in metabolic pathways could account for these effects. The importance of this research is emphasized in light of ongoing concerns about bacterial resistance to antibiotics.

Shcheglov (2002) examined MMW on *e coli* cells at various cell densities and frequencies. His work suggests that cell-to-cell communication may be involved in bacterial responses to weak EMF.

Bulgakova (1996) irradiated staphylococcus cultures with different frequencies of MMW with non-thermal intensities with short exposure periods (minutes). He found changes in bacterial sensitivities developed in 5 of 14 antibiotics used in sublethal concentrations with both suppression and stimulation of growth.

Research on 5G millimeter waves

This section provides an assortment of 5G millimeter wave studies as a way to consolidate research previously presented in the first part of the book, as well as additional studies related specifically to the implementation of 5G

5G millimeter wavelengths reportedly can have an effect through heat (tissue destruction) through a resonant effect of increased vibration in an object the size of the wavelengths, and at low power levels through signaling of skin structures that can affect metabolism, the nervous system, the endocrine system, and the reproductive system. Millimeter wavelengths at high intensity have also been used in military applications in Active Denial Systems (9 MMW for non-lethal crowd control weapons) that create heat to repel persons. Studies show these systems can burn the skin. (Declassified Military Studies).

An older Russian paper, "Biological Effects of Millimeter Wavelengths" by Zalyuboxskaya (1977) was declassified by the CIA. This paper describes the research on both humans and animals showing a myriad of adverse effects of millimeter wavelengths. The author notes that millimeter wave technology had been used for years without any studies on biological effects. The researchers found that "millimeter waves caused changes in the body manifested in structural alterations in the skin and internal organs, qualitative and quantitative changes in the blood and bone marrow composition, and changes in the conditioned reflex activity, tissue respiration . . . and nuclear metabolism. The degree of unfavorable effect of millimeter waves depended on the duration of radiation and individual characteristics of the organism."

The author confirmed that millimeter waves do not penetrate skin but act on nerve receptors in the skin to cause such diverse biological and metabolic effects as a reduction in hemoglobin and erythrocytes, higher blood cortisol levels, adrenal stimulation, mitochondrial dysfunction and suppression of the central nervous system with notable changes in liver, kidneys, heart, and brain.

"Current state and implications of research on biological effects of millimeter waves, a review of the literature" (Pakhomov 1998) reviewed dozens of research findings on low-intensity millimeter waves and concluded that the reported MMW effects "could not be readily explained by temperature changes during irradiation." The review concludes by questioning the adequacy of regulatory limits, stating that, "Safety limits for these types of exposure are based solely on predictions of energy deposition and MMW heating, but in view of current studies, this approach is not necessarily adequate."

In "Skin Heating and Injury by Prolonged Millimeter-Wave Exposure: Theory Based on a Skin Model Coupled to a Whole Body Model and Local Biochemical Release From Cells at Supraphysiologic Temperatures" (Stewart 2006), the authors write ". . .the biophysical mechanism of biochemical release through cell membranes within the tissue regions that reach supraphysiologic (greater than amounts normally found in the body) temperatures is also considered. The released molecules are delivered to other skin regions by diffusion and into the bloodstream by perfusion, where according to our hypothesis, the molecules interact with susceptible cells. This raises the possibility of additional indirect injury at nearby deeper skin regions that experience insignificant heating. Biochemical release may also lead to injury at distant sites within the body by perfusion clearance that transfers molecules into the systemic circulation to reach other susceptible cells."

A review of the research in 2010 noted that "A large number of cellular studies have indicated the millimeter wavelengths (MMW) may alter structural and functional properties of membranes." Exposure to MMWs may affect the plasma membrane either by modifying ion channel activity or by modifying the phospholipid bilayer. Water molecules also seem to play a role in these effects. Skin nerve endings are a likely target of MMWs and the possible starting point of numerous biological effects. MMWs may activate the immune system through stimulation of the peripheral neural system (Ramundo-Orlando 2010).

In humans, the penetration depth of more than 90% of the transmitted power is absorbed in the epidermal and dermal layers (Wu 2015b). Because the depth is so superficial, higher heating occurs more quickly with less dissipation. Many biological responses to MMW irradiation can be initiated with the skin (Isaac 2012; Ziskin 2013; Gandhi and Riazi 1986).

While the cellular industry claims there is no danger with 5G because the waves are only superficially absorbed by the skin of humans, it has been shown that 4G and low band 5G emissions can cause oxidation of tissues (Yakymenko 2016).

In a 2019 review study, "5G Wireless Communication and Health Effects," by Simko and Mattsson (2019), 94 *in vitro* and *in vivo* millimeter wave studies were examined. There were no epidemiological studies found to review. It is noted that this study was funded by a grant from Deutsche Telecom, the largest telecom provider in Europe and one of the largest in the world.

The authors state, "The available studies do not provide adequate and sufficient information for a meaningful safety assessment, or for the question about non-thermal effects. There is a need for research regarding local heat developments on small surfaces, e.g., skin or the eye, and on any environmental impact."

They conclude, "In summary, the majority of studies with MMW exposures show biological responses. . . . . Regarding the quality of the presented studies, too few studies fulfill the minimal quality criteria to allow any further conclusions."

Cell Antennas increase exposures in communities

A published review of effects of Wi-Fi radiation entitled, "Wi-Fi is an important threat to human health," (Pall 2018a) found that repeated Wi-Fi studies show that Wi-Fi causes oxidative stress, sperm/testicular damage, neuropsychiatric effects including EEG changes, apoptosis (cell death), cellular DNA damage, endocrine changes, and calcium overload.

"The impact of radiofrequency radiation on DNA damage and antioxidants in peripheral blood lymphocytes of humans residing in the vicinity of mobile phone base station" (Zothansiama 2017), is a research study that compared people living close (within 80 meters) and far (more than 300 meters away) from cellular antennas. It found that the people living closer had several significant changes in their blood predictive of cancer development. An earlier 2016 study found DNA damage in peripheral blood lymphocytes (Gulati 2016).

"Mortality by neoplasia and cellular telephone base stations," is a 10 year study by the Belo Horizonte Brazil Health Department and several universities in Brazil that found an elevated relative risk of cancer mortality at residential distances of 500 meters or less from cell installations (Dode 2011). Shortly after this study was published, the city prosecutor sued several cell phone companies and requested that almost half of the city's antennas be removed. Many antennas were dismantled.

A 2019 study, "RF-EMF exposure induced by mobile phones operating in 4G LTE (Long Term Evolution) small cells in two different urban cities (Mazloum 2019), found increased RF-EMF exposure from small cell LTE networks in France and the Netherlands. Researchers measured the RF-EMF from LTE, MC (macro cells, meaning large cell towers) and SC networks (low-powered small cell base stations) and found that the small cell networks increased the radio emissions from base stations (called downlink) a factor of 7-46 while decreasing the radio emissions from user equipment exposure (a factor of 5-17).

So while the cellular devices themselves could emit less radiation, the cell antennas will increase the levels from cell antennas. This study shows the increased exposures would be involuntary. We can turn our phone off, but we cannot turn off the antennas in the neighborhood.

A 2018 multi-country study, "Comparison of radiofrequency electromagnetic field exposure levels in different everyday microenvironment in an international context," measured RF in several countries and found that cell phone tower radiation is the dominant contributor to RF exposure in most outdoor areas. Exposure in urban areas was higher and exposure has drastically increased. As an example, the measurements the researchers took in Los Angeles, USA, were 70 times higher than the US EPA estimate 40 years ago (Sagar 2018).

An Australian study found that children in kindergartens with nearby antenna installations had nearly three-and-a-half times higher RF exposures than children with installations further away (more than 300 meters) (Bhatt 2017).

Another study looked at 529 children in Denmark, the Netherlands, Slovenia, Switzerland and Spain who wore meters around the waist or carried in a backpack during the day and placed close to the bed at night. Researchers found "the largest contributors to total personal environmental RF-EMF exposure were downlink (meaning from cell tower base stations) and broadcast" (Birks 2018).

Impact of RF-EMR

A 2018 article, "Planetary electromagnetic pollution: it is time to assess its impact" (Bandara 2018), points to unprecedented increasing RF exposures. The authors write, "Due to the exponential increase in the use of wireless personal communication devices (eg., mobile or cordless phones and Wi-Fi or Bluetooth-enabled devices) and the infrastructure facilitating them, levels of exposure to radiofrequency electromagnetic radiation around the 1 GHz frequency band, which is mostly used for modern wireless communications, have increased from extremely low natural levels by about 10 to the 18th power.. . . It is plausibly the most rapidly increasing anthropogenic [originating in human activity] environmental exposure since the mid-20th century, and levels will surge considerably again, as technologies like the Internet of Things and 5G add millions more radiofrequency transmitters around us."

Abdel-Rassoul (2007), in "Neuobehavioral effects among inhabitants around mobile phone base stations," found that living nearby mobile phone base stations (cell antennas) increased the risk for neuropsychiatric problems such as headaches, memory problems, dizziness, tremors, depression, sleep problems and some changes in the performance of neurobehavioral functions.

A 2010 landmark review, "Biological Effects from Exposure to Electromagnetic Radiation Emitted by Cell Tower Base Stations and Other Antenna Arrays" (Levitt and Lai 2010), reviewed 100 studies and found approximately 80% showed biological effects near towers. They write, "Both anecdotal reports and some epidemiology studies have found headaches, skin rashes, sleep disturbances, depression, decreased libido, increased rates of suicide, concentration problems, dizziness, memory changes, increased risk of cancer, tremors, and other neurophysiological effects in populations near base stations."

Carpenter (2013) in "Human disease resulting from exposure to electromagnet fields," summarized the evidence, stating that excessive exposure to magnetic fields from power lines and other sources of electric current increases the risk of development of some cancers and neurodegenerative diseases, and that excessive exposure to RF radiation increases risk of cancer, male infertility, and neurobehavioral abnormalities.

In "Changes of Neurochemically Important Transmitters under the Influence of Modulated RF Fields - A Long Term Study Under Real Life Conditions," Bucher and Eger (2011) show elevated levels of stress hormones (adrenaline, noradrenaline), and lowered dopamine and PEA levels in urine in area residents during the first six months of cell tower installation. Even after 1.5 years, the levels did not return to normal.

Martin L. Pall, in "Microwave electromagnetic fields act by activating voltage-gated calcium channels: why the current international safety standards do not predict biological hazard," writes "It can be seen from the above that ten different well-documented microwave EMF effects can be easily explained as being a consequence of EMF VGCC (voltage-gated calcium channels) activation: oxidative stress, elevated single and double strand breaks in DNA, therapeutic responses to such EMFs, breakdown of the blood-brain barrier, cancer, melatonin loss, sleep dysfunction, male infertility and female infertility."

Belpomme (2018) in "Thermal and non-thermal health effects of low intensity non-ionizing radiation: An international perspective," finds that exposure to low frequency and radiofrequency electromagnetic fields at low intensities poses a significant health hazard that has not been adequately addressed by national and international organizations such as the World Health Organization. There is strong evidence that excessive exposure to mobile phone-frequencies over long periods of time increases the risk of brain cancer both in humans and animals. The mechanism(s) responsible include induction of reactive oxygen species, gene expression alteration and DNA damage through both epigenetic and genetic processes. *In vivo* and *in vitro* studies demonstrate adverse effects on male and female reproduction, almost certainly due to generation of reactive oxygen species (ROS).

Smart Meters

Smart meters are used to monitor electric, gas, and water services in some areas. The smart meter is a wireless two-way transmitter that pulses signals to communicate continuously between a customer's home and the utility company. They have also been placed in schools and workplaces.

The wireless smart meter typically produces relatively potent and very short pulsed RF/microwaves, emitting these millisecond-long RF burst on average of 9,600 times a day. According to testimony by the California Utility Pacific Gas & Electric, before that state's Public Utilities Commission, a peak level emission is two and a half times higher than the stated safety signal (California State Public Utilities Commission).

Unfortunately, the Federal Communications Commission (FCC) radio frequency radiation safety standards are for short term thermal effects measured at 5 and 30 minutes and are configured for a 6'2", 200-pound male. Further, the meters are assessed for safety compliance in isolation, not in a mesh network, in which they are designed to operate, or in combination with multiple meters and other radio frequency (RF) sources.

A wireless smart meter produces radio frequency microwave radiation with two antennas in approximately the same frequency range (900 MHz to 2.4 GHz) as a typical cell tower. Depending on how close it is to occupied space within a home, a smart meter can cause much higher radio frequency exposures than cell towers commonly do.

It is reported that if a smart meter is located on a common wall with a bedroom or kitchen rather than a garage wall, for example, the RF exposure can be the same as being with 200 to 600 feet distance of a cell tower with multiple carriers. With both cell towers and smart meters, the entire body is immersed by microwaves that go out in all directions, which increases the risk of overexposure to many sensitive organs such as the eyes, testicles and heart. In contrast, with a cell phone, people are exposed to microwaves primarily in the head and neck unless they are using speaker mode, and only when the device is turned on or in standby mode.

The figures for radio frequency (RF) exposure given by utilities are time-averaged numbers which hide the peak power of the smart meter and disguise the fairly continuous nature of the pulses. Smart meters are unlike cell phones or Wi-Fi in their pattern of sharp spikes of RF.

The meter is part of a nationwide project dubbed Advanced Metering Infrastructure (AMI). This is often referred to as the "smart grid." The AMI smart meter records electrical consumption data and sends the information wirelessly to energy system managers, in this case, your utility company. It can be programmed to read and transmit data monthly, or up to every fifteen seconds.

The AMI came about through the American Recovery and Investment Act of 2009. Part of the objectives of the Act were to provide temporary relief programs for those most affected by the recession and invest in infrastructure, education, health, and renewable energy. The federal government has put an estimated eleven billion dollars into incentive programs for utilities that participate.

In other words, this is not simply your local utility company hooking you up so your information can be read at any time and reported back to you so you can see what your current use is via an online program. It is a nation-wide federal program backed by the federal government for utilities.

The AMI's purpose is to expose all Americans to three new and powerful sources of microwave radiation. They are the "smart" meter, "smart" appliances, and the ubiquitous network of antennas on utility poles and cell towers in urban and rural neighborhood that will power the rollout of 5G.

Neither the federal government nor those profiteering from this rollout have undertaken a single public health study about the long-term health effects of exposure to electromagnetic radiation (EMR) from "smart" meters. This is in spite of the medical literature now loaded with peer-reviewed studies about the non-thermal biological effects of exposure to EMR.

Peer reviewed studies report DNA damage, abnormal genetic and hormonal changes, sperm damage, pregnancy complications, weakening of the blood-brain barrier, disturbance of voltage-gated calcium channels (VGCC) (for example in the heart), degradation of immunity, and certain types of cancers.

Especially worrisome is mounting evidence that inescapable electromagnetic fields, such as exposure from smart meters, places children at particular risk for altered brain development and for impaired learning and behavior. These concerns are corroborated by the BioInitiative Report (2012).

The report was produced by twenty-nine medical and public health experts from ten countries and offers a meta-analysis of over eighteen hundred new scientific studies showing that chronic exposure to both electric field exposure and microwaves poses a serious health hazard.

Can you opt out of having a smart meter? In 2009, FERC, the Federal Energy Regulatory Commission, which regulates, monitors and investigates electricity, natural gas, hydropower and such, issued a Policy Statement on Smart Grid Policy that acknowledged the Energy Independence and Security Act of 2007. The act aims to move the US towards greater energy independence and security and increase the efficiency of products, buildings and vehicles.

However, the Act gave FERC no new authority to enforce such standards. While energy bills in 2005 and 2007 were codified into public laws, no part of them creates a federal law pertaining to individual consumers or dictating that the public must be forced to comply with provisions of the smart grid.

But opting out is really not a solution. Even if a homeowner is allowed to opt out, they are still involuntarily irradiated 24/7 by their neighbors' smart meters. This source of radio frequency radiation, unlike a cell phone which can be turned off or placed at a distance from your body, is out of an individual's control (Global Indoor Health Network https://www.globalindoorhealthnetwork.com/smart-meters).

The Environment

Dr. Cindy Russell published a remarkable study in 2018 entitled, "Wireless Silent Spring" in the Santa Clara County Medical Association Bulletin. It paralleled the seminal work of Rachel Carson in her 1962 book "Silent Spring", a publication widely credited with ushering in the modern environmental movement (Carson 1962). Carson's book contains the story of the annihilation of birds, squirrels, fish, earthworms and beneficial insect after the introduction of ever more toxic pesticides.

Russell examines the similarities between the silent spring created in cities and farms from pesticides and that of wireless technology and the rapid spread of cell towers. She examines the effects of this technology that biologists have found on wildlife and then compares the histories, mechanisms and impacts between pesticides and wireless radiation (Russell 2018b).

Wireless radio frequency affects navigations of birds and bees. Biologists have discovered that wireless radio frequency radiation (RFR) disturbs internal magneto-receptors they use for orientation. In addition, non-ionizing radiation can have profound impacts on the natural environment by disruption of other complex cellular and biologic processes in mammals, birds, fish, amphibians, insects, trees, plants, seeds and bacteria.

Reported adverse effects from radio frequency radiation that have been identified include abnormal behavior, developmental abnormalities, diminished reproduction and increased mortality. The effects of this radiation may not be immediately apparent with a slow decline in the health of wildlife seen over time with cumulative exposure, adding a new environmental toxin contributing to silent springs in cities, orchards and farms.

The more towers being built, the more additive mix of radiation frequencies saturate the environment. This creates a toxic air space. However, the FCC does not consider non-thermal (non-heat related) biological effects in its current guidelines, in spite of studies which show results in this area. Appropriate safety testing and regulation of this technology is lacking, but the invention, commercialization and deployment of cell towers, and now 5G "small cells," continues.

Biologists have discovered that birds' magnetic compass orientation appears vulnerable to weak broadband electromagnetic fields (Pakhomov 2017; Schwarze 2016; Wiltschko 2015). A german scientist, Svenja Engels (2014) led the research project to confirm this effect over a period of seven years before publishing the data.

Studies on bees have found behavioral effects (Kumar 2011), disrupted navigation (Goldsworthy 2009; Sainudeen 2011; Kimmel 2007), decreasing egg laying rate (Sharma and Kumar 2010), and reduced colony strength (Sharma and Kumar 2010; Harst 2006).

Researchers are now attributing wireless radiation from cellular communications to be a significant contributing cause of bee "colony collapse disorder," insect disappearance, and the decline in house sparrows in London (Balmori 2007; Everaert 2007). Currently forty percent of bird species are under critical threat. Insects are not only important pollinators, they are the base of the food chain for birds, amphibians, reptiles and mammals.

Bees are a critical pollinator species for agricultural productivity (Schwagerl 2016). Of the 100 crops that provide 90% of the world's food supply, 71 are pollinated by bees, according to the UN environmental Program (UN #Friday Fact 21). Bee numbers have plummeted in the last two decades. Contributing factors affecting the health and reproduction of bees include pesticides, global climate change, loss of habitat and air pollution, with new research pointing towards microwave radiation as an important and yet unrecognized cause for concern.

Electromagnetic microwave radiation has been shown to disrupt bee behavior. Bees have demonstrated aggression after 30 minutes of cell phone exposure (Favre 2011). A cell phone placed next to a beehive appears to cause a slow destruction of the hive (Dallo 2015).

Lazaro (2016) looked at the effect of mobile communication antennas on the abundance and composition of wild pollinators on two Greek islands with variable distances from cell towers, carefully measuring the radio frequency radiation. He found negatives in all pollinators except butterflies.

Belgian entomologist Marie-Claire Camaerts (2017) found that insects are particularly sensitive to RFR. Other research found bees may be particularly affected by manmade electromagnetism (UN #Friday Fact, CNN Business Insider) When crossing such electromagnet fields, bees may no longer remember their way, may no longer fly in the correct direction, and may become unable to return to their hive.

Russell writes, "These are truly alarming findings and serve as a dire warning on further wireless expansion, especially with regards to sensitive wildlife areas and agricultural rural zones that depend on pollination" (Russell 2018b).

Proposed 5G millimeter wavelengths (MMW) are a similar size to insects and this creates a damaging vibrational effect on the organism known as resonance. An example of resonance is that of a wineglass which shatters when an opera star reaches a high note, vibrating air molecules matching the glasses' natural oscillating frequency.

In general, mechanical resonance occurs when the frequency of an oscillation matches the system's or its subcomponent's natural frequency and this results in increasingly intensified additive vibration with more energy being absorbed, causing more disturbance of the system.

At low power, an effect is greatly magnified. Thielens (2018) looked at this effect on four different insects exposed to electromagnet fields from 2 to 120 GHz. He noted, "The insects show a maximum in absorbed radio frequency power at wavelengths that are comparable to their body size. . .This could lead to changes in insect behaviour, physiology, and morphology over time due to an increase in body temperatures, from dielectric heating."

In addition, a newer technology is incorporated into these 5G systems called phased arrays. These powerful "beam steering" arrays scan back and forth from tower to device for easier connection with an individual's movement, to detect the device, similar to the surface-to-air missile systems used in military radar. They're also used in AM and FM Broadcast stations and planned for automotive sensors and satellites. It is unknown what effect this increase in power and density of environmental radiation will have on our beneficial insects and pollinators.

Several review studies point to wildlife harm (Balmori 2015; Cucarachi 2012; MOE 2010). Cucarachi reported that reviewing 113 peer-reviewed publication revealed, "In about two thirds of the reviewed studies, ecological effects of RF-EMF was reported at high as well as at low dosages. The very low dosages are compatible with real field situations, and could be found under environmental conditions."

Trees are being negatively impacted by RF radiation. An experiment on trembling Aspen trees points to ambient electromagnetic radiation from a variety of sources (cell towers, satellites, RF from electric power generation) causing poor growth and smaller leaves. There was no impact on shielded trees (Haggerty 2010).

Waldman-Selsam (2016) conducted a 9-year study showing cell tower radiation caused the death of nearby trees over time. The study involved over 100 trees and found trees sustained more damage on the side of the tree facing the antenna.

Russell tallies many parallels between the silent spring of pesticides and wireless radiation. Both are invisible, working on a cellular level which may slowly cause disease. Both are universal in our environment, with the continuous pulsating waves of radiation straying into any nearby living organism, be it human, pet or wildlife.

Exposure to both are from cradle to grave, with a host of wireless devices in the home, and cell towers which are projected to be implemented every 300 meters (around 1000 feet) in order to accommodate 5G. Like pesticides sprayed in large areas to kill a few flying insects, wireless radiation is sprayed in all directions in order to find intended devices, penetrating all living organisms and causing cellular damage and affecting our ecosystem (Balmori 2010; Cucurachi 2012; Sivani and Saravanamuttu 2013; NTP 2018).

Balmori (2009) presents a review on the impact of radio   frequency   radiation   from   wireless telecommunications on wildlife. He writes "Electromagnetic radiation is a form of environmental pollution which may hurt wildlife. Phone masts located in their living areas are irradiating continuously some species that could suffer long-term effects, like reduction of their natural defenses, deterioration of their health, problems in reproduction and reduction of their useful territory through habitat deterioration." Electromagnetic radiation can exert an aversive behavioral response in rats, bats and birds such as sparrows. He concludes "microwave and radiofrequency pollution constitutes a potential cause for the decline of animal populations and deterioration of health of plants living near phone masts."

Both pesticides and wireless radiation cause a variety of adverse biological effects. 2G radiation was found to cause DNA damage and an increase in the risk of cancer of the heart, brain and adrenal medulla in a recent 10-year, $25-million National Toxicology Program study (NTP 2018). Non-ionizing radiation from 3G and 4G cell towers have been found to cause nonspecific symptoms of electrosensitivity in some living within 300 meters of a cell tower including insomnia, dizziness, brain fog, fatigue, depression and heart palpitations. Cell phone radiation has been associated with harm to the reproductive system, neurologic system, immune system and hematologic system (BioInitiative Report 2014; Oceana Report).

Both are children born of war. Pesticides were first developed as agents of chemical warfare. Radiofrequency microwave technology was developed in World War II as radar. At the end of the war, microwave ovens were developed (Ackerman 2016). Millimeter technology (95 GHz) has been developed for crowd control called the Active Denial System (Active Denial System).

Both are bio-toxic. Wireless radio frequency radiation has been shown to have a primary mechanism of harm from oxidation. Yamenko (2016) looked at 100 studies of RF radiation both *in vivo* and *in vitro* and found 93 showed oxidation as a mechanism of toxicity.

Wireless technology has continued to evolve and expand. While 5G is being introduced, the old systems—2G, 3G, 4G—will still be in place. With the proposed 5G technology and the Internet of Things (IoT), industry aims to integrate this with other wireless generations, and even open up any remaining radio frequency spectrums, creating a blanket of mixed frequency wireless radiation that wildlife and humans will be exposed to.

Radiation emissions are not only from cell towers, but also in remotely controlled stratospheric balloons (Loon Project), far orbiting satellites and proposed low orbiting satellites which will greatly increase ambient levels of electromagnetic radio frequency radiation (EMR).

Like pesticides, there has been inadequate research examining the mix of frequencies we are exposed to. The 2018 NTP study, which found clear evidence of carcinogenicity as well as DNA damage and cardiomyopathy, looked only at 2G wireless technology.

There are no government plans for testing of 3G, 4G, or 5G individually or in combination. Testing for synergistic effects of wireless radiation and toxic chemicals has also not been attempted.

Despite a virtual research vacuum on 5G high frequency radiation, federal and state legislation is being introduced and quickly approved to ensure the rapid deployment of this technology by removing local jurisdiction and limiting fees for cities and counties to use the public right of way.

Pesticides appear to be more toxic to people who do not have the metabolic pathways to transform and excrete them. Wireless radio frequency radiation is observed to cause non-specific symptoms of headaches, dizziness, insomnia, nausea, irritability, depression and heart palpitations in those who are electrosensitive. Classic symptoms of electrosensitivity also occur in a high number of those living near cell towers and when a cell tower is removed, symptoms resolve (Santini 2002; Navarro 2003; Shinjyo 2014).

Belpomme (2015) conducted a large clinical study and found laboratory biomarkers that connect multiple chemical sensitivity to electrosensitivity. It also has been noted that having these conditions causes predictable isolation and fear which can lead to neuropsychiatric symptoms (Egerle 2016).

Both the pesticide industry and the cellular industry are deceptive. David Michaels in "Doubt is Their Product" quotes a tobacco executive saying, "Doubt is our product since it is the best means of competing with the "body of fact" that exists in the minds of the general public. It is also the means of establishing a controversy."

Pesticides have been well protected by the industry that created them. An investigation of over 20,000 documents including internal scientific studies, meeting minutes and memos from federal regulatory agencies and manufacturers was led by the Center for Media and Democracy and the Bioscience Research Project resulted in "The Poison Papers" (2018). Concealment, political manipulation cover-up and collusion were found along with suppression of fraudulent independent research and secrecy of the toxic effects of chemicals and pesticides.

Wireless telecommunications have been regulated by the Federal Communications Commission (FCC) since the 1996 Telecommunications Act was passed. The Environmental Protection Agency (EPA) was relieved of their oversight duty of radio frequency radiation just prior to that.

The 1996 Act assumed, even before testing, that there were no health or environmental effects of this radiation. It is specified in the law that health and environmental effects cannot be used as an argument to deny cell tower placement. This has hampered attempts to monitor or identify health effects in the United States.

Harvard's Center for Ethic's publication called "Captured Agency: How the Federal Communications Industry is Dominated by the Industries it Presumably Regulated" (Alster 2015) highlights the industries exorbitant lobbying influence to the tune of about $400 million a year, according to the Center for Responsive Politics. A revolving door in Washington was also noted with telecom industry executives filing the critical "independent" government positions.

In her excellent book, "Disconnect: The Truth About Cell Phone Radiation, What the Industry Has Done to Hide It, and How to Protect Your Family" Dr. Devra Davis documents industry manipulation along with their discrediting of scientists who have identified and published literature on the adverse health effects of wireless radiation (Davis 2013).

Dr. Russell concludes her publication of "Wireless Silent Spring: by saying, "We are just beginning to understand the fragile biologic complexities of the Earth's living creatures as we simultaneously document natures decline under the dismissing hand of mankind. Many have warned that our fate will follow that of nature."

"The expansion of wireless technologies for human convenience will require more cell towers on every street corner. This will threaten natural ecosystems in favor of immersive and invasive technology which is contributing to both negative environmental, physical and mental health effects, especially on our youth."

"Instead of increasing the number of cell towers, we need to be removing cell towers near schools, homes, businesses and hospitals as well as in wildlife areas." Alternatives such as fiberoptic networks and cable exist that are faster, more fire resistant, use less energy and are cheaper in the long run (NIS 2018)."

"Traditional copper landlines are reliable in emergencies, cheap, already built and connect everyone without risk. . . . We can have the benefits of faster, dependable and more private communications without compromising public or environmental health." (Russell 2018b)

Conclusion

As it stands right now, you can take steps to decrease electromagnetic fields from your cell phone by enhancing your distance from it. And you can hopefully choose not to live next to a 2G-3G-4G cellular tower. But if 5G is implemented in your area, and small cells are installed on the utility poles outside your home or workplace, you have no way to shield yourself from the pulsing dangers of 5G millimeter waves.

The bottom line is that we need a moratorium on 5G and any further wireless antenna deployment until potential hazards for human health and the environment have been fully investigated by scientists independent from the wireless industry. And this is the exact message sent to President Trump in December of 2019, by a broad coalition of scientists, doctors and advocates.

According to the National 5G Resolution letter:

- The FCC has stated that 800,000 antenna sites will be required to fully deploy 5G in the United States, with global deployments expected to reach almost 5 million by 2021.
- New wireless antennas are rapidly being installed on streetlights and utility poles directly in front of homes and schools.
- 5G will dramatically increase the general public's daily exposure to radiofrequency electromagnetic fields (RF-EMF), in addition to the emissions from 2G, 3G, 4G wireless infrastructure already in place.
- 5G was not premarket safety tested.

- Research shows biological effects from levels of wireless exposures presently allowed.
- Research on 5G emissions shows serious impacts to humans, bees, trees and wildlife.

The scientific studies noted in this book can be summarized in the findings of Martin L Pall, PhD, Professor of Biochemistry and Basic Medical Sciences at Washington State University (Pall 2018b).

Non-thermal microwave frequency EMF exposures:

- Attack our nervous systems including our brains, leading to widespread neurological/neuropsychiatric effects and possibly many other effects.
- Attack our endocrine, our hormonal, systems. The consequences of the disruption of these two regulatory systems is immense.
- Produce oxidative stress and free radical damage, which have central roles in essentially all chronic diseases.
- Attack the DNA of our cells, producing single strand and double strand breaks in cellular mutations in cells which produce mutations in future generations.
- Produce elevated levels of apoptosis (programmed cell death), events especially important in causing both neurodegenerative diseases and infertility.
- Lower male and female fertility, lower sex hormones, lower libido and increased levels produce excessive intracellular calcium and excessive calcium signaling.
- Attack the cells of our bodies to cause cancer. Such attacks are thought to act via 15 different mechanisms during cancer causation.

- Cause other effects including cardiac effects, very early onset dementias, and possibly ADHD and autism.

The FCC and EPA must be held accountable to the American people. They must take back control from the telecommunications industry, rather than allowing the industry to set regulatory and technical standards on their own.

There is a noted lack of funding in the US for independent scientific research on health effects of radio frequency electromagnet radiation (RF-EMR) that is free of industry influence or bias. Who will provide this funding, given obstacles of a cultural, economic and political nature?

We are busy people and we love our cells phones. MMW and RF-EMR are invisible and detrimental effects probably won't be seen for many years. Who will take up the challenge?

References

5G Appeal. Welcome to the 5G Appeal.eu page. www.5gappealeu/about/ 2017

Abdel-Rassooul G, El-Fateh OA, Salem MA, Michael A, Farahat F, El-Batanouny M, Salem E. Neurobehavioral effects among inhabitants around mobile phone base stations. Neurotoxicology 2007 Mar; 28 (2): 434-40

Ackerman, Evan. A Brief History of the Microwave Oven: Where the "radar" in Raytheon's Radarange came from. IEEE Spectrum 30 Sept 2016

Active Denial System FAQs. Non-Lethal Weapons Program. https://jnlwp.defense.gov/About/Frequently-Asked-Questions/Active-Denial-System-FAQs/

ACS: American Cancer Society 2019 https://www.cancer.org/cancer/cancer-causes/radiation-exposure/cellular-phones.html

Adams, J.A., Galloway, T.S., Mondal, D., Esteves, S.C., Mathews, F., 2014. Effect of mobile telephones on sperm quality: a systematic review and meta-analysis. Environ. Int. 70, 106–112. ⟨http://www.ncbi.nlm.nih.gov/pubmed/24927498

Adebayo EA, Adeeyo AO, Ogundiran MA, Olabisi O. Bio-physical effects radiofrequeny electromagnetic radiation (RF-EMR) on blood parameters, spermatozoa, liver, kidney and heart of albino rate. <u>Journal of King Saud University – Science</u> Vol 31 Issue 4, October 2019, Pages 813-821

Agarwal, A., Deepinder, F., Sharma, R.K., Ranga, G., Li, J., 2008. Effect of cell phone usage on semen analysis in men attending infertility clinic: an observational study. <u>Fertil. Steril.</u> 2008 (Jan), 89. ⟨http://www.ncbi.nlm.nih.gov/pubmed/17482179-⟩.

Agarwal A, Desai NR, Makker K, Varghese A, Mouradi R, Sabanegh E, Sharma R. Effects of radiofrequency electromagnetic waves (RF-EMW) from cellular phones on human ejaculated semen: an in vitro pilot study. <u>Fertil Steril</u> 2009 Oct; 92(4): 1318-25

Agarwal, A., Singh, A., Hamada, A., Kesari, K., 2011. Cell phones and male infertility: a review of recent innovations in technology and consequences. <u>Int. Braz. J. Urol.</u> 37 (4), 432–454. ⟨http://www.ncbi.nlm.nih.gov/pubmed/21888695⟩.

Al-Ali BM, Paatzak J, Fischereder K, Pummer K, Shamloul R. Cell Phone Ussage and Erectile Function. <u>Cent European J Urol</u> 2013 66(1) 75-7

Alexiou GA, Sioka C. Mobile phone use and risk for intracranial tumors. <u>J Negat Results Biomed</u> 2015; 14; 23

Alfadda AA, Sallam RM. Reactive Oxygen Species in Health and Disease. J Biomed Biotechnol 2012 936486 2012

Al-Quzwini, O.F., Al-Taee, H.A., Al-Shaikh, S.F., 2016. Male fertility and its association with occupational and mobile phone towers hazards: an analytic study. Middle East Fertil. Soc. J. 21 (4), 236–240. ⟨http://www.sciencedirect.com/science/article/pii/ S11105690163001127

Anderson LE. Biological Effects of Extremely Low-Frequency Electromagnetic Fields: In Vivo Studies. American Industrial Hygiene Association Journal Vol 54 1993 Iss 4

Balmori A, Electromagnetic pollution from phone masts. Effects on wildlife. Pathophysiology Vol 16, Issues 2-3, August 2009, Pages 191-199

Balmori A, Hallberg O. The Urban Decline of the House Sparrow (Passer domesticus): A Possible Link with Electromagnetic Radiation. (2007)

Balmori A. Anthropogenic radiofrequency electromagnetic fields as an emerging threat to wildlife orientation. Science of the Total Environment Vol 518-519, 15 June 2015, Pages 58-60

Bandara P, Weller. Cardiovascular disease: Time to identify emerging environmental risk factors. <u>European Journal of Preventative Cardiology</u> October 3, 2017

Bandara P, Carpenter DO. Planetary electromagnetic pollution: it is time to assess its impact. <u>The Lancet Planetary Health</u> Vol 2, Issue 12, Dec 2018, Pages e512-d514

Banik, S., 2003. Bioeffects of microwave—a brief review. <u>Bioresour. Technol.</u> 87, 155–159. ⟨https://www.researchgate.net/profile/Dr_Shyamal_Banik/publication/ 10745209_Bioeffects_of_microwave_-A_brief_review/links/ 543f95d80cf27832ae8b8e41.pdf⟩.

Belpomme D, Campagnac C, Irigaray P. Reliable disease biomarkers characterizing and identifying electrohypersensitivity and multiple chemical sensitivity as two etiopathogenic aspects of a unique pathological disorder. <u>Rev Environ Health</u> 2015;30(4):251-71

Belpomme D, Hardell L, Belyaev I, Burgio E, Carpenter DO. Thermal and non-thermal health effects of low intensity non-ionizing radiation: An international perspective. <u>Environ Pollut</u> 2018 Nov;24(Pt A):643-658

Belyaev, Igor. Non-thermal biological effects of microwaves. <u>Microw. Rev.</u> 2005 11, 13–29. ⟨http://www.tandfonline.com/doi/abs/10.1080/153683 70500381844? journalCode=iebm20⟩.

Belyaev IY, Markova E, Hillert L, Malmgren LO, Persson BR. Microwaves from UMTS/GSM mobile phones induce long-lasting inhibition of 53BP1/gamma-H2aX DNA repair foci in human lymphocytes. Bioelectromagnetics 2009 Feb;30(2):129-4

Benson VS, Pirie K, Schuz J, Reeves GK, Beral V, Green J, Million Women Study Collaborators. Mobile phone use and risk of brain neoplasms and other cancers: prospective study. Int J Epidemiol 2013 Jun;42(3):792-802

Betzalel, N, Feldman Y, Ishal PB. The Modeling of the Absorbance of Sub-THz Radiation by Human Skin. IEEE Transactions on Terahertz Science and Technology Vol 7, Issue: 5, Sept. 2017

Betzalel N, Ishai B, Feldman Y. The human skin as a sub-THz receiver – Does 5G pose a danger to it or not? Environ Res 2018 May;163:208-216

Bhatt CR, Redmayne M, Billah B, Abramson MJ, Benke G. Radiofrequency-electromagnetic field exposures in kindergarten children. J Expo Sci Environ Epidemiol 2017 Sep;27(5):497-504

BioInitiative 2012. The BioIinitiative Report 2012. A Rational for Biologically-based Public Exposure Standards for Electromagnetic Fields (ELF and RF) https://bioinitiative.org 2019

Birks, L., Guxens, M., Papadopoulou, E., Alexander, J., Ballester, F., et al., 2017. Maternal cell phone use during pregnancy and child behavior problems in five birth cohorts. International Society for environmental epidemiology. Environ. Health Perspect. http://dx.doi.org/10.1016/j.envint.2017.03.024. ⟨https://www.ncbi.nlm.nih.gov/ pubmed/28392066

Birks LE, Struchen B, Eeftens M, van Wei L, Huss A, Gajsek P, Kheifets L, Gallastegi M, et al. Spatial and temporal variability of personal environmental exposure to radio frequency electromagnetic fields in children in Europe. Environ Int 2018 Aug 117:204-214

Blank M, Goodman R. Electromagnetic Fields Stress Living Cells. Pathophysiology 2009 Aug 16 (2-3) 71-8

Blank, M., Havas, M., Kelly, E., Lai, H., Moskowitz, J., 2015. International appeal: scientists call for protection from non-ionizing electromagnetic field exposure. Eur. J. Oncol. 20 (3–4). ⟨http://mattioli1885journals.com/index.php/ Europeanjournalofoncology/article/view/4971⟩.

Bormusov E, Andley UP, Sharon N, Schächter L, Lahav A, et al. ,2008. Non-thermal electromagnetic radiation damage to lens epithelium. Open Ophthalmol. J. 2, 102–106. ⟨https://www.ncbi.nlm.nih.gov/pmc/articles/PMC2694 600/⟩

Bortkiewicz A, Gadzicka E, Szymczak W. Mobile phone use and risk for intracranial tumors and salivary gland tumors – A meta-analysis. <u>Int J Occup Med Environ Health</u> 2017 Feb 21;30(1):27-43

Buldak RJ, Polaniak, Buldak L, Zwirska-Korczala K, Skonieczna M, Monsiol A, Kukla M, Dulawa-Buldak A, Birkner E. Short-term Exposure to 50 Hz ELF-EMF Alters the Cisplatin-Induced Oxidative Response in AT478 Murine Squamous Cell Carcinoma Cells. <u>Bioelectromagnetics</u> 33 (8); 641-51

Bulgakova, V.G., Grushina, V.A., Orlova, T.I., Petrykina, Z.M., Polin, A.N., Noks, P.P., Kononenko, A.A., Rubin, A.B., 1996. Effect of millimeter-band radiation of nonthermal intensity on the sensitivity of staphylococcus to various antibiotics. <u>Biofizika</u> 41, 1289–1293. (in Russian).
⟨https://www.ncbi.nlm.nih.gov/pubmed/9044624⟩.

California State Public Utilities Commission 2011. Pacific Gas and Electric Company's response to administrative law judge's October 18, 2011 ruling directing it to file clarifying radio frequency information.

Cammaerts MC. Is electromagnetism one of the causes of the CCD? A work plan for testing this hypothesis. <u>J of Behavior</u> 2(1):1006. Mar 28, 2017

Cardis E, Armstrong BK, Bowman JD, Giles GG, Hours M, Krewski D, McBride M, Parent ME, et al. Risk of brain tumours in relation to estimated RF dose from mobile phones: results from five Interphone countries. <u>Occup Environ Med</u> 2011 Sept; 68(9):631-640

Carlberg M, Hardell L. Pooled analysis of Swedish case-control studies during 1997-2003 and 2007-2009 on meningioma risk associated with the use of mobile and cordless phones. <u>Oncol Rep</u> 2015 Jun;33(6):3093-8

Carlberg M, Hedendahl L, Ahonen M, Koppel T, Hardell L. Increasing incidence of thyroid cancer in the Nordic countries with main focus on Swedish data. <u>BMC Cancer</u> 2016;16: 426

Carlberg M, Hardell L. Evaluation of Mobile Phone and Cordless Phone Use and Glioma Risk Using the Bradford Hill Viewpoints from 1965 on Association or Causation. <u>Biomed</u> 2017; 2017 9218486

Carpenter DO. Extremely low frequency electromagnetic fields and cancer: How source of funding affects results. <u>Environmental Research</u> Vol 178, Nov 2019, 108688

Carpenter DO. Human disease resulting from exposure to electromagnetic fields. 2013;28(4):159-72

Carpenter, RL Van Ummersen, CA., 1968. The action of microwave radiation on the eye. <u>J. Microw. Power</u> 3 (1), 3–

19. http://dx.doi.org/10.1080/00222739.1968. 11688664. ⟨http://www.tandfonline.com/doi/abs/10.1080/002227 39.1968. 11688664⟩

Carson, Rachel.   Silent Spring. 1962. Houton Mifflin Company Boston

Castello PR, Hill I, Sivo F, Portelli L, Barnes F, Usselman R, Martino CF. Inhibition of cellular proliferation and enhancement of hydrogen peroxide production in fibrosarcoma cell line by weak radio frequency magnetic fields. Bioelectromagnetics 2014 Dec;35(8)598-602

Catterall WA. Voltage-Gated Calcium Channels. Cold Spring Harb Perspect Biol 2011 Aug; 3(8): a003947

CDC.   2015   Vision   Health   Initiative. ⟨https://www.cdc.gov/visionhealth/basics/ced/ index.html⟩.

CDC 2019. Centers for Disease Control and Prevention: Reproductive   Health https://www.cdc.gov/reproductivehealth/infertility/inde x.htm

Chapman S, Azizi L, Luo Q, Sitas F. Has the incidence of brain cancer risen in Australia since the introduction of mobile phones 29 years ago? Cancel Epidemiol 2016 Jun;42:199-205

Chemakov GM, Korochkin VL. Reactions of biological systems of various complexity to the action of low-level EHF radiation. <u>Radioelectronica Moscow</u> 1989

Christensen HC, Schuz J, Kosteljanetz M, Poulsen HS, Boice JD Jr, McLaughlin JK, Johansen C. Cellular telephones and risk for brain tumors: a population-based, incident case-control study. <u>Neurology</u> 2005 Apr 12;64(7):1189-95

Consales, C.,2012.Electromagnetic fields, oxidative stress ,and neurodegeneration. <u>Int.J. Cell Biol.</u> 2012 (2012). http://dx.doi.org/10.1155/2012/683897

Coureau G, Bouvier G, Lebailly P, Fabbro-Peray P, Gruber A, Leffondre K, Guillamo JS, Louiseau H, Maathoulin-Pelissier S, Salamon R, Baldi I. Mobile phone use and brain tumours in the CERENAT case-control study. <u>Occup Environ Med</u> 2014 Jul;71(7):514-22

Cucurachi S et al. A review of the ecological effects of radiofrequency electromagnetic fields (EMF). 2012 <u>Environment International</u> 51C:116-140RF

Cucurachi S, Tarnis WLM, Vijver MG, Peijnenburt, WJGM, Bolte JFB, de Snoo GR. A review of the ecological effects of radiofrequency electromagnetic fields (RF-EMF). <u>Env Int</u> Vol 51, Jan 2013, Pages 116-140

Cutz A. Effects of microwave radiation on the eye: the occupational health perspective. <u>Lens and Eye Toxicity Research</u> 31 Dec 1988 6(1-2):379-86

D'Andrea, J.A., Chalfin, S., 2000. Effects of microwave and millimeter wave radiation on the eye. In: Klauenberg, B.J., Miklavčič, D. (Eds.), Radio Frequency Radiation Dosimetry and Its Relationship to the Biological Effects of Electromagnetic Fields. NATO Science Series

Dasdag, S., Akdag, M.Z., Erdal, M.E., Erdal, N., Ay, O.I., Ay, M.E., Yilmaz, S.G., Tasdelen, B., Yegin, K., Effects of 2.4 GHz radiofrequency radiation emitted from Wi-Fi equipment on microRNA expression in brain tissue. Int. <u>J. Radiat. Biol</u>. 2015. 91 (7), 555–561. http://dx.doi.org/10.3109/09553002.2015.1028599

Davis DL, Kesare S, Soskolne CL, Miller AB, Stein Y. Swedish review strengthens grounds for concluding that radiation from cellular and cordless phones is a probably human carcinogen. <u>Pathophysiology</u> 2013 Apr;20(2):123-9

Davis, Devra. Disconnect: The Truth about Cell Phone Radiation, What the Industry Has Done to Hide It and How to Protect Your Family. www.Amazon.com 2013

Declassified Military Studies. Kiev Vrachebnoye Delo in Russian No 3. Biological Effect of Millimeter Radiowaves , 1977 pp 116-119 Declassified and approved for release 2012

Delo, Kiev Vrachebnoye. Russian No 3. Biological Effect of Millimeter Radiowaves , 1977 pp 116-119 Declassified and approved for release 2012

Desai NR, Kesari KK, Agarwal A. Pathophysiology of cell phone radiation: oxidative stress and carcinogenesis with focus on male reproductive system. https://www.ncbi.nim.nih.gov/pmc/articles/PMC2776019/

Di Ciaula A. Towards 5G communication systems: Are there health implications? Int J Hyg Environ Health 2018 Apr;221(3):367-375

Dode AC, Leao M, Tejo FDAF, Gomes ACR, Dode DC, Dode MC, Moreira CW, Condessa VA, Albinatti C, Caiaffa WR. Mortality by neoplasia and cellular telephone base stations in the Belo Horizonte municipality, Minas Gerais state, Brazil. Science of the Total Environment Vol 409, Iss 19, 1 Sep 2011, Pages 3649-3665

Dréan, Y.L., Mahamoud, Y.S., Page, Y.L., Habauzit, D., Le Quément, C., Zhadobov, M., Ronan Sauleau, R., 2013. State of knowledge on biological effects at 40–60GHz. Comptes Rendus Phys. 14 (5), 402–411. http://dx.doi.org/10.1016/j.crhy.2013.02.005. (http://www.sciencedirect.com/science/article/pii/S1631070513000480).

Duan W et al. Comparison of the Genotoxic Effects Induced by 50 Hz Extremely Low-Frequency Electromagnetic Fields and 1800 MHz Radiofrequency Electromagnetic Fields in GC-2 Cells. Radiation Research Mar 2015 Vol 183 No 3 pp 305-314

Eghlidospour M, Mortazavi SMJ, Yousefi F, Mortazavi SAR. New Horizons in Enhancing the Proliferation and Differentiation of Neural Stem Cells Using Stimulatory Effects of the Short Time Exposure to Radiofrequency Radiation J Biomed Phys Eng 2015 Sep;5(3):95-104

El-Gohary OA, Said MA. Effect of electromagnetic waves from mobile phone on immune status of male rats: possible protective role of vitamin D. Can J Physiol Pharmacol 2017 Feb;95(2):151-156

Eltiti S, Wallace D, Zougkou K, Russo R, Joseph S, Rasor P, Fox eE. Development and Evaluation of the Electromagnetic Hypersensitivity Questionaire. Bio Electro Magnetics 2007 28:137-151

Emas O. Effects of electromagnetic field exposure on the heart: a systematic review. Toxicology and Industrial Health 2016 Vol 32(1) 76-82

Engels S, Schneirr NL, Lefeldt N, Hein CM, Zapka M, Michalik A, Elbers D, Kittel A, Hore PJ, Mouritsen H. Anthropogenic electromagnetic noise disrupts magnetic compass orientation in a migratory bird. Nature 2014 May 15:509(7500):353-6

Erdal N, Gurgul S, Tamer L, Avaz L. Effects of long-term exposure of extremely low frequency magnetic field on oxidative/nitrosative stress in rat liver. J Radiat Res 2008 Mar; 49(2):181-7

Erogul O, Oztas E, Yidirim, Kir T, Avdur E, Komesli G, Irkilata HC, Irmak MK, Peker AF. Effects of electromagnetic radiation from a cellular phone on human sperm motility: an in vitro study. Arch Med Res 2006 Oct;37(7):840-3

EPA, 1981. Index of Publications on Biologic Effects of Electromagnetic Radiation (0–100GHz) 1981. List 3627 Studies to 1980.

EPA, 1992. Electric and Magnetic Fields: An EPA perspective on research needs and priorities for improving health risk assessment

EPA, 1993. US EPA Office of Air and Radiation and Office of Research and Development: Summary of Results of the April 26–27, 1993. Radiofrequency Radiation conference. Anaylsis of Panel Discussions. Volume 1. March 1995

Esmekaya, M.A., Ozer, C., Seyhan, N., 2011. 900MHz pulse-modulated radiofrequency radiation induces oxidative stress on heart, lung, testis and liver tissues. Gen. Physiol. Biophys. 30 (1), 84–89.

http://dx.doi.org/10.4149/gpb_2011_01_84.
⟨https://www. ncbi.nlm.nih.gov/pubmed/21460416⟩.

Everaert J, Bauwens D.   A possible effect of electromagnetic radiation from mobile phone base stations on the number of breeding house sparrows (Passer domesticus). <u>Electromagn Biol Med</u> 2007;261(1):63-72

Falcioni L, Bua L, Tibaldi E, Lauriola M, De Angelis L, Gnudi F, Mandrioli D, Manservigi M, Manservisi F et al. Report of final results regarding brain and heart tumors in Sprague-Dawley rats exposed from prenatal life until natural death to mobile phone radiofrequency field representative of a 1.8 GHz GSM base station environmental emission. <u>Environ Res</u> 2018 Aug;165:496-503

Favre D. Mobile phone-induced honeybee worker piping. <u>Apidologie</u> May 2011, Vol 42, Iss 3, pp 270-279

Favre Daniel.   Disturbing Honeybees' Behavior with Electromagnetic Waves: a Methodology. 2017 <u>J of Behavior</u> 2(2) 1010

FCC, 1996. Telecommunications Act of 1996. ⟨https://www.fcc.gov/general/ telecommunications-act-1996⟩.

FCC, 1997a. Guidelines for Evaluating the Effects of Environmental Radiofrequency Radiation) ET Docket No. 93-62

FCC, 1997b. Current guidelines – evaluating compliance with FCC Guidelines for human exposure to radiofrequency electromagnetic fields. OET Bull (65 Edition 97-01. August 1997). ⟨http://citeseerx.ist.psu.edu/viewdoc/download?doi=10. 1.1.159. 3824&rep=rep1&type=pdf⟩.

FCC, 2013. Radio Frequency Safety. ⟨https://www.fcc.gov/general/radio-frequencysafety-0⟩. FCC, 2015. RF Safety FAQ. Updated November 25, 2015. ⟨https://www.fcc.gov/ engineering-technology/electromagnetic-compatibility-division/radio-frequencysafety/faq/rf-safety#Q24⟩.

FCC: FCC Fact Sheet. Accelerating Wireline Broadband Deployment by Removing Barriers to Infrastructure Investment. Sep 5, 2018

FCC Letter 5G Americas, 2016. FCC Filing 14-177 5G Americas. Posted 10/3/16. ⟨https://www.fcc.gov/ecfs/filing/1093077796003/ document/1093077796003462f⟩.

Fernandez C, de Salles AA, Sears ME, Morris rD, Davis DL. Absorption of wireless radiation in the child versus adult brain and eye from cell phone conversation or virtual reality. <u>Environ Res</u> 2018 Nov;167:694-699

Foster, M.R., Ferri, E.S., Hagan, G.J., 1986. Dosimetric study of microwave cataractogenesis. <u>Bioelectromagnetics</u> 7 (2), 129–140. ⟨https://www.ncbi.nlm.nih.gov/pubmed/3741488⟩.

Franzellitti S, Valbonesi P, Ciancaglini N, Biondi C, Contin A, Bersani F, Fabbri E. Transient DNA damage induced by high-frequency electromagnetic fields (GSM 1.8 GHz) in the human trophoblast HTR-8/SVneo cell line evaluated with the alkaline comet assay. <u>Mutation Research/Fundamental and Molecular Mechanisms of Mutagenesis</u> Jan 2010 Vol 68f3, Iss 1-2:35-42

Frei, P, Poulsen AH, Johansen C, Olsen JH, Steding-Jessen M, Schuz J. Use of mobile phones and risk of brain tumours: update of Danish cohort study. <u>BMJ</u> 2011 343:d6387

Frentzel-Beyme, R.,. John R. Goldsmith on the usefulness of epidemiological data to identify links between point sources of radiation and disease. <u>Public Health Rev</u>. 1994 22 (3–4), 305–320. ⟨https://www.ncbi.nlm.nih.gov/pubmed/7708942⟩.

Fung, Brian. Court deals blow to FCC's bid to speed 5G rollout. <u>CNN Business</u> Aug 9, 2019 <u>https://edition.cnn.com/2019/08/09/tech/5g-fcc-regulations-ruling/index.html</u>

Gandhi, O.P., Riazi, A., 1986. Absorption of millimeter waves by human beings and its biological implications.

IEEE Trans. Microw. Theory Tech. MTT 34 (2), 228–235. ⟨http://ieeexplore.ieee.org/document/1133316/⟩.

Gapeev, A.B., Lushnikov, K.V., Shumilina, Iu.V., Sirota, N.P., Sadovnikov, V.B., Chemeris, N.K., 2003. Effects of low-intensity extremely high frequency electromagnetic radiation on chromatin structure of lymphoid cells in vivo and in vitro. Radiats Biol. Radioecol. (1), 87–92. ⟨https://www.ncbi.nlm.nih.gov/pubmed/12677665⟩.

Gee, D., 2009. Late lessons from early warnings: towards realism and precaution with EMF? Pathophysiology 16 (2–3), 217–231. http://dx.doi.org/10.1016/j.pathophys.2009.01.004. ⟨https://www.ncbi.nlm.nih.gov/labs/articles/19467848/⟩.

Giuliani L, Soffritti M. Non-Thermal Effects and Mechanisms of Interaction Between Electromagnetic Fields and Living Matter: An ICEMS Monograph. Ramazzini Institute Eur J Oncol Library Vol 5 2010

Glotova KU, Sadchikova MN. Development and clinical course of cardiovascular changes after chronic exposure to microwave irradiation, JPRS 51238. Arlington VA Joint Publications Research Service 25 Aug 1970 3 pp

Goldsworthy, Andrew. The Birds, the Bees and Electromagnetic Pollution. https://ecfsapi.fcc.gov/file/7520958012.pdf May 2009

Grigoriev, Y.G., Mikhailov, V.F., Ivanov, A.A., Maltsev, V.N., Ulanova, A.M., StavrakovaI, N.M., Nikolaeva, A., Grigoriev, O.A., 2010. Autoimmune processes after long-term low-level exposure to electromagnetic fields part 4. oxidative intracellular stress response to the long-term rat exposure to nonthermal RF EMF. Biophysics 55, 1054–1058. ⟨https://link.springer.com/article/10.1134/S0006350910 060308⟩.

Gulati S, Yadav A, Kumar N, Kanupriva, NK, Kumar R, Gupta R. Effect of GSTM1 and GSTT1 Polymorphisms on Genetic Damage in Humans Populations Exposed to Radiation from Mobile Towers. Arch Environ Contam Toxicol 2016 Apr;70(3):615-25

Gustavs, Katherine. Changes of Clinically Important Neurotransmitters under the Influence of Modulated RF Fields (Rimbach Study). EMFacts Consultancy 4 Sep 11

Haggarty, Katie. Adverse influence of radio frequency background on trembling aspen seedlings: Preliminary observations. International Journal of Forestry Res. 2010 Apr (6)

Hardell L, Hallquist A, Mild KH, Carlberg M, Pahlson A, Lilja A. Cellular and cordless telephones and the risk for brain tumours. Eur J Cancer Prev 2002 Aug;22(4):377-86

Hardell L, Carlberg M, Hannsson Mild K. Pooled analysis of two case-control studies on use of cellular and cordless telephones and the risk for malignant brain tumours diagnosed in 1997-2003. Int Arch Occup Environ Health 2006 Sep;79(8):630-9

Hardell L, Carlberg M, Hansson Mild K. Epidemiological evidence for an association between use of wireless phones and tumor diseases. Pathophysiology 2009 Aug 16(2-3):113-22

Hardell L, Carlbert M, Hansson Mild K. Mobile phone use and the risk for malignant brain tumors: a case-control study on decreased cases and controls. Neuroepidemiology 2010 Aug;35(2):109-14

Hardell L, Carlberg M, Hansson Mild K. Pooled analysis of case-coantrol studies on malignant brain tumours and the use of mobile and cordless phones including living and deceased subjects. Int J Oncol 2011 May 38(5):1465-74

Hardell L, (2013a) Carlberg M, Soderqvist F, Kjell Hansson Mild. Case-control study of the association between malignant brain tumours diagnosed between 2007 and 2009 and mobile and cordless phone use. Int J Oncol 2013 Dec; 43(6):1833-1845

Hardell L, (2013b) Carlbert M, Hansson Mild K. Use of mobile phones and cordless phones is associated with increased risk for glioma and acoustic neuroma. Pathophysiology 2013 Apr 20(2):85-110

Hardell L, Carlberg M.Mobile phone and cordless phone use and the risk for glioma – Analys of pooled case-control studies in Sweden, 1997-2003 and 2007-2009. Pathophysiology 2014 Mar 22(1):1-13

Hardell L. World Health Organization, radiofrequency radiation and health – a hard nut to crack (Review). International Journal of Oncology June 21, 2017, pages 405-413

Hardell L, Carlberg C. Comments on the US National Toxicology Program technical reports on toxicology and carcinogenesis study in rats exposed to whole-body radiofrequency radiation at 900 MHz and in mice exposed to whole-body radiofrequency radiation at 1,900 MHz. Int J Oncol 2019 Jan; 54(1):111-127

Harst W. Kuhn J, Stever H. Can Electromagnetic Exposure Cause a Change in Behavior? Studying Possible Non-Thermal Influences on Honey Bees—An Approach with the Framework of Educational Informatics. http://www.next-up.org/pdf/ICRW_Kuhn_Landau_study.pdf

Hassanshahi, A., Shafeie, S.A., Fatemi, I., Hassanshahi, E., Allahtavakoli, M., Shabani, M., Roohbakhsh, A., Shamsizadeh, A.,2017. The effect ofWi-Fi electromagnetic waves in unimodal and multimodal object recognition tasks in male rats. Neurol. Sci. 38 (6), 1069–1076.

http://dx.doi.org/10.1007/s10072-017-2920-y. ⟨https://www.ncbi. nlm.nih.gov/pubmed/28332042⟩.

Havas M, Marrongelle J, Pollner B, Kelley E, Rees CRG, Tully L. Provocation Study using Heart Rate Variabiity shows Radiation from 2.4 GHz Cordless Phone affects Autonomic Nervous System. <u>European Journal of Oncology</u> Jan 2010

Havas M. Radiation from wireless technology affects the blood, the heart, and the autonomic nervous system. <u>Rev Environ Health</u> 2013; 28(2-3):75-84

Havas M. When theory and observation collide: Can non-ionizing radiation cause cancer? <u>Environ Pollut</u> 2017 Feb;221:501-505

Hojo, S., Tokiya, M., Mizuki, M., Miyata, M., Kanatani, K.T., 2016. Development and evaluation of an electromagnetic hypersensitivity questionnaire for Japanese people. Bioelectromagnetics 37 (6), 353–372. http://dx.doi.org/10.1002/bem.21987. ⟨https://www.ncbi.nlm.nih.gov/pmc/articles/PMC5094 565/⟩.

Huss, A., Egger, M., Hug, K., Huwiler-Müntener, K., Röösli, M., 2007. Source of funding and results of studies of health effects of mobile phone use: systematic review of experimental studies. Environ. Health Perspect. 115 (1),

14.⟨https://www.ncbi.nlm.
nih.gov/pmc/articles/PMC1797826/⟩.

ICNIRP, 2009. International Commission on Non-Ionizing Radiation Protection: ICNIRP statement on the 'Guidelines for limiting exposure to time-varying electric, magnetic, and electromagnetic fields (up to 300 GHz)'. Health Phys. 97, 257–258. ⟨https://www.ncbi.nlm.nih.gov/pubmed/19667809⟩.

INTERPHONE Study Group 2010. Brain tumour risk in relation to mobile telephone use: results of the INTERPHONE international case-control study. Int J Epidemiol 2010 Jun; 39(3):675-94

INTERPHONE Study Group 2011. Acoustic neuroma risk in relation to mobile telephone use: results of the INTERPHONE international case-control study. Cancer Epidemiol 2011 Oct;35(5):453-64

Isaac, M., Chiu,1,2,3 Christian, A., von Hehn,1,2,3, Woolf, Clifford J., 2012. Neurogenic inflammation – the peripheral nervous system's role in host defense and immunopathology. Nat. Neurosci. 15 (8), 1063–1067. http://dx.doi.org/10.1038/nn. 3144. ⟨https://www.ncbi.nlm.nih.gov/pmc/articles/PMC3520 068/⟩.

Jensen, CF. Legal Opinion on whether it would be in contravention of human rights and environmental law to establish the 5G-system in Denmark. Bonnor Advokater 2019

Johansen C, Boice J Jr, McLaughlin J, Olsen J. Cellular telephones and cancer—a nationwide cohort study in Denmark. J Natl Cancer Inst 2001 Feb 7;93(3):203-7

Kesari KK, Kumar S, Behari J. 900-MHz microwave radiation promotes oxidation in rat brain. Electromagn Biol Med 2011 Dec;30(4):219-334

Kesari KK (2012a), Behari J. Evidence for Mobile Phone Radiation Exposure Effects on Reproductive Pattern of Male Rats: Role of Ros. Electromagnetic Biology and Medicine 31:3, 213-222

Kesari KK (2012b), Kumar S, Behari J. Pathophysiology of Microwave Radiation: Effect on Rat Brain. Appl Biochem Biotechnol Jan 2012 166(2): 379-88

Kesari KK, Agarwal A, Henkel R. Radiations and male fertility. Reprod Biol Endocrinol 2018 Dec 9;16(1):118

Kim, J.H.,Yu, D.H., Huh,Y.H., Lee, E.H.,Kim, H.G., Kim,H.R., 2017. Long-term exposure to 835MHz RF-EMF induces hyperactivity, autophagy and demyelination in the cortical neurons of mice. Sci. Rep. 20 (7), 41129. http://dx.doi.org/10.1038/ srep41129. ⟨https://www.nature.com/articles/srep41129⟩.

Kimmel S, Kuhn J, Harst W, Stever H. Electromagnetic Radiation: Influences on Honeybees (*Apis mellifera*) (2007) https://www.researchgate.net/publication/292405747_E lectromagnetic_radiation_Influences_on_honeybees_Api s_mellifera_IIAS-InterSymp_Conference

Kivrak EG, Yurt KK, Kaplan AA, Alkan I, Altun G. Effects of electromagnetic fields exposure on the antioxidant defense system. J Microsc Ultrastruct 2017 Oct-Dec; 5(4): 167-176

Klaeboe L, Blaasaas KG. Tynes T. Use of mobile phones in Norway and risk of intracranial tumours. Eur J Cancer Prev 2007 Apr:16(2):158-64

Kolomytseva, M.P., Gapeev, A.B., Sadovnikov, V.B., Chemeris, N.K., 2002. Suppression of nonspecific resistance of the body under the effect of extremely high frequency electromagnetic radiation of low intensity. Biofizika 47 (1), 71–77. ⟨https://www. ncbi.nlm.nih.gov/pubmed/11855293⟩.

Kostoff RN. Adverse Effects of Wireless Radiation. Ch 4. Kostoff and Lau, 2017. 2019. PDF. http://hdl.handle.net/1853/61946

Koyu A, Ozguner F, Yilmaz Hr, Uz E, Cesur G, Ozcelik N. The Protective Effect of Caffeic Acid Phenethyl Ester (cAPE) on Oxidative Stress in Rat Liver Exposed to the

900 MHz Electromagnetic Field. <u>Toxicol Ind Health</u> 25 (6) 429-34

Koziorowska A, Waszkiewicz EM, Romerowicz-Misielak M, Zglejc-Waszak K, Franczak A. Extremely low-frequency electromagnetic field (EMF) generates alterations in the synthesis and secretion of oestradiol-1uB(E2) in uterine tissues: An in vitro study. <u>Theriogenology</u> 2018 Apr 1;110:86-95

Kues, H.A., Monhan, J.C., 1992. Microwave-induced changes to the primate eye. Johns. Hopkins <u>APL Tech.</u> Dig. 13 (1) (1992, PDF Needed). ⟨http://www.jhuapl.edu/ techdigest/views/pdfs/V13_N1_1992/V13_N1_1992_K ues.pdf⟩.

Kumar NR, Sangwan S, Badotra P. Exposure to cell phone radiations produces biochemical changes in worker honey bees. <u>Toxicol Int</u> 2011 Jan-Jun; 18(1): 79-72

Kumar S, Behari J, Sisodia R. Influence of electromagnetic fields on reproductive system of male rats. <u>Int J Radiat Biol</u> 2013 Mar, 89(3): 147-54

Kumar S, Nirala JP, Behari J, Paulraj R. Effect of electromagnetic irradiation produced by 3G mobile phone on male rat reproductive system in a simulated scenario. <u>Indian J Exp Biol</u> 2014 Sep;52(9):890-7

LA Times Editorial Board. Editorial: An audacious 5G power (pole) grab July 5, 2017 https://www.latimes.com/opinion/editorials/la-ed-power-pole-grab-20170705-story.html

Lahkola A, Salminen T, Raitanen J, Heinavaara, Schoemaker MJ, Christensen HC, Feychting M, Johansen C, et al. Meningioma and mobile phone use—a collaborative case-control study in five North European countries. Inj J Epidemiol 2008 Dec;37(6):1304-13

Lai H, Singh NP. Melatonin and N-tert-butyl-alpha-phenylnitrone Block 60pHz Magnetic Field-Induced DNA Single and Double Strand Breaks in Rat Brain Cells. J Pineal Res 22 (3) 152-62 Apr 1997

Lai H, Singh NP. Magnetic-field-induced DNA strand breaks in brain cells of the rat. Environ Health Peerspect 2004; 112: 687-694

Lasalvia M, Scrima R, Perna G, Piccoli C, Capitanio N, Biagi PF, Schiavulli L, Ligonzo T et al. Expposure to 1.8 GHz electromagnetic fields affects morphology, DNA-related Raman spectra and mitochondrial functions in human lympho-monocytes. PLoS One 2018 Feb 20;13(2):e0192894

La Vignera, S., Condorelli, R.A., Vicari, E.,D'Agata, R.,Calogero, A.E.,2012. Effects of the exposure to mobile phones on male reproduction: a review of the literature. J.

Androl. 33 (3), 350–356. ⟨http://www.ncbi.nlm.nih.gov/pubmed/21799142⟩.

Lazaro, Amparo. Electromagnetic radiation of mobile telecommunication antennas affects the abundance and composition of wild pollinators. J of Insect Conservation 2016 Apr 20(2):1-10

Lerchi A, Klose M, Grote K, Wilhelm AFX, Spathmann O, Fiedler T, Streckert J, Hansen V, Clemens M. Tumor promoton by exposure to radiofrequency electromagnetic fields below exposure limits for humans. Biochemical and Biophysical Research Communications Vol 459, Iss 4, 17 Apr 2015, 585-590

Leszczynski, Dariusz. Seminar on the 5G and Health: Gaps in the Knowledge. Georges River, NSW, Australia, Sep 15, 2019.

Levine, H., Jørgensen, N., Martino-Andrade, A., Mendiola, J., Weksler-Derri, D., Mindlis, I., Pinotti, R., Swan, S.H., 2017. Temporal trends in sperm count: a systematic review and meta-regression analysis. 25 July 2017. ⟨https://academic.oup.com/humupd/ article/doi/10.1093/humupd/dmx022/4035689/Tempor al-trends-in-sperm-counta-systematic-review⟩.

Levitt BB, Lai H. Biological effects from exposure to electromagnetic radiation emitted by cell tower base stations and other antenna arrays. NRC Research Press Aug 2010

Lipman, R.M., Tripathi, B.J., Tripathi, R.C., 1988. Cataracts induced by microwave and ionizing radiation. <u>Ophthalmol</u> 33 (3), 200–210. 〈https://www.ncbi.nlm.nih.gov/pubmed/3068822〉.

Liu, K., Li, Y., Zhang, G., Liu, J., Cao, J., Ao, L., Zhang, S., 2014. Association between mobile phone use and semen quality: asystemic review and meta-analysis. <u>Andrology</u> 2 (4), 491–501. 〈https://www.ncbi.nlm.nih.gov/pubmed/24700791〉.

Lonn S, Ahibom, Hall P, Feychting M, Swedish Interphone Study Group. Long-term mobile phone use and brain tumor risk. <u>Am J Epidemiol</u> 22005 Mar 15;161(6):526-35

Lushnikov, K.V., Gapeedv, A.V., Shumilina, Iu.V., Shibaev, N.V., Sadovnikov, V.B., Chmeris, N.K., 2003. Decrease in the intensity of the cellular immune response and nonspecific in flammation upon exposure to extremely high frequency electromagnetic radiation. Biofizika 48 (5), 918–925. 〈https://www.ncbi.nlm.nih.gov/pubmed/14582420〉.

Lymar SV, Khairutdinov RF, Hurst JK. Hydroxyl radical formation by O-O bond homolysis in peroxynitrous acid. <u>Inorg Chem</u> 2003 Aug 25;42(17):5259-66

Magras IN, Xenos TD. RF radiation-induced changes in the prenatal development of mice. <u>Bioelectromagnetics</u> 1997;18(6):455-61

Mashevich M, Folkman D, Kesar A, Barbul A, Korenstein R, Jerby E, Avivi L. Exposure of human peripheral blood lymphocytes to electromagnetic fields associated with cellular phones leads to chromosomal instability. <u>Bioelectromagnetics</u> Vol 24 Iss 2 Jan 2003

Mazloum T, Aerts S, Joseph W, Wiart Joe. RF-EMF exposure induced by mobile phones operating in LTE small cells in two different urban cities. <u>Annals of Telecommunications</u> 2019 Vol 74,pg 35-42

Mazor R, Korenstein-Ilan A, Barbul A, Eshet Y, Shahadi A, Jerby E, Korenstein R. Increased Levels of Numeical Chromosome Aberrations After in Vitro Exposure of Human Peripheral Blood Lymphocytes to Radiofrequency Electromagnetic Fields for 72 Hours. <u>Radiat Res</u> 169 (1) 28-37 Jan 2008

Megha K, Deshmukh PS, Banerjee BD, Tripathi AK, Abegaonkar MP. Microwave Radiation Induced Oxidative Stress, Cognitive Impairment and Inflammation in Brain of Fischer Rats. <u>Indian J Exp Biol</u> 50(12) 889-96

Michaels, David. Doubt is Their Product. 2008. Oxford University Press

Michelozzi P, Capon A, Kirchmayer U, Forastiere F, Biggeri An, Barca A, Perucci CA. Adult and Childhood Leukemia near a High-Power Radio Station in Rome, Italy. <u>American Journal of Epidemiology</u> Vol 155, Iss 12, 15 June 2002, pg 1096-1103

Mihai CT, Rotinberg P, Brinza F, Vochita G. Extremely low-frequency electromagnetic fields cause DNA strand breaks in normal cells. <u>J of Env Health Sci & Eng</u> 2014: 12:15

Miller AB, Morgan LL, Udasin I, Davis DL. Cancer epidemiology update, following the 2011 IARC evaluation of radiofrequency electromagnetic fiels (<u>Monograph</u> 102). <u>Env Res</u> Vol 167 Nov 2018, pg 673-683

MOE. The Ministry of Environment and Forest. Report on Possible Impacts of Communication Cell Towers on Wildlife Including Birds and Bees.

Morgan LL, Kesare S, David DL. Why children absorb more microwave radiation than adults: The consequences. <u>J Microscopy and Ultrastructure</u> Vol 2, Iss 4, Dec 2014, pg 197-204 2010

Morgan, L.L., 2009. Estimating the risk of brain tumors from cellphone use: published case- control studies. <u>Pathophysiology</u>   16   (2–3),   137–147. ⟨http://www.ncbi.nlm. nih.gov/pubmed/19356911⟩.

Morgan LL, Miller AB, Sasco A, David DL. Mobile phone radiation causes brain tumors and should be classified as a probable human carcinogen (2A) (review). <u>Int J Oncol</u> 2015 May;46(5):1865-71

Mortazavi SAR, Mortazavi G, Mortazavi SMJ. Comments on "Radiofrequency electromagnetic fields and some cancers of unknown etiology: An ecological study". <u>Sci Total Environ</u> 2017 Dec 31: 609:1

Mortazavi SMJ. Exposure Level Matters! Exposure to Radiofrequency Electromagnetic Fields Produced by Mobile Phones and Brain Cancer. Sep 29, 2017 <u>https://www.linkedin.com/pulse/exposure-level-matters-radiofrequency-electromagnetic-smj-mortazavi/?published=t</u>

Mortazavi SMJ, Mortazavi SAR, Haghani M. Evaluation of the validity of a Nonlinear J-Shaped Dose-Response Relationship in Cancers Induced by Exposure to Radiofrequency Electromagnetic Fields. <u>J Biomed Phys Eng</u> 2019 Aug; 9(4):487-494

Moss, C.E., 1997. Report of electromagnetic radiation surveys of video display terminals. NIOSH. CDC. ⟨https://www.cdc.gov/niosh/nioshtic-2/00081009.html⟩.

Myung, S.K., Ju, W., McDonnell, D.D., Lee, Y.J., Kazinets, G., Cheng, C.T., Moskowitz, J.M., 2009. Mobile phone use and risk of tumors: a meta-analysis (2009) mobile phone use and risk of tumors: a meta-analysis. <u>J. Clin. Oncol.</u> 20 (33), 5565–5572. (Nov 20, 2009.)(Published online first Oct 13, 2009.). 〈http://ascopubs.org/doi/full/10.1200/jco.2008.21.6366〉.

Naarala J, Kesari KK, McClure I, Chavarriaga C, Juutilainen J, Martino CF. Direction-Dependent Effects of Combined Static and ELF Magnetic Fields on Cell Proliferation and Superoxide Radical Production. <u>Biomed Res Int</u> 2017;2017:5675086

Nasim I, Kim S. Human Exposure to RF Fields in 5G Downlink. 10 Nov 2017 https://arxiv.org/pdf/1711.03683.pdf

National 5G Resolution Letter 2019. Dozens of US Doctors and Healthcare Practitioners Send Letter to President Trump Calling for a Moratorium on 5G. https://ehtrust.org/dozens-of-us-doctors-and-healthcare-practitioners-send-letter-to-president-trump-calling-for-a-moratorium-on-5g-press-release/

Navarro EA et al. The Microwave Syndrome: A Preliminary Study in Spain. 2003 <u>Electrobiology and Medicine</u> Dec 2002

Neufeld E, Kuster N. Systematic Derivation of Safety Limits for Time-Varying 5G Radiofrequency Exposure Based on Analytical Models and Thermal Dose. <u>Health Phys</u> 2018 Sep 21

Nielsen C, Hui R, Lui WY, Solovyov LA. Towards predicting intracellular radiofrequency radiation effects. <u>PLoS One</u> March 14, 2019

Nguyen TH, Shamis Y, Croft RJ, Wood A, McIntosh RL, Crawford RJ, Ivanova EP. 18 GHz electromagnetic field induces permeability of Gram-positive cocci. <u>Sci Rep</u> Jun 16;5:10980

NIH NEH, 2010. National Eye Institute. National Institutes of Health. Prevalence of Cataracts 2010. ⟨https://nei.nih.gov/eyedata/cataract/tables⟩.

Nittby H. Brun A, Stromblad S, Moghadam MK, Sun W, Malmgren L, Eberhardt J, Persson BR, Salford LG. Nonthermal GSM RF and ELF EMR effects upon rat BBB permeability. <u>The Environmentalist</u> June 2011, Vol 31, Iss 2, pp 140-148

NTP. National Toxicology Program. NTP Technical Report on the Toxicology and Carcinogenesis Studies in Hsd: Sprague Dawley SD Rats Exposed to Whole-Body Radio Frequency Radiation at a Frequency (900 MHz) and Modulations (GSM and CDMA) Used by Cell Phones. National Institutes of Health, Public Health Service. US Department of Health and Human Services.

Oceana Radiofrequency Scientific Advisory Association. ORSAA https://www.orsaa.org/orsaa-database.html

Odaci E, Ozyilmaz C.   Exposure to a 900 MHz electromagnetic field for 1 hour a day over 30 days does change the histopathology and biochemistry of the rat testis. <u>Int J Radiat Biol</u> 2015 Jul;01(7): 547-54

Oh JJ, Byun SS Lee SE, Choe G, Hong SK. Effect of Electromagnetic Waves from Mobile Phones on Spermatogenesis in the Era of 5G-LTE. <u>Biomed Res Int</u> 2018 Jan 29;2018:1801798

Othman, H., Ammari, M., Sakly, M., Abdelmelek, H., 2017a. Effects of prenatal exposure to WIFI signal (2.45GHz) on postnatal development and and behavior in rat: influence of maternal restraint. <u>Behav. Brain Res.</u> 326, 291–302. http://dx.doi.org/10. 1016/j.bbr.2017.03.011. 〈https://www.ncbi.nlm.nih.gov/pubmed/28288806〉.

Ozmen I, Naziroglu M, Alici HA, Sahin F, Cengiz M, Eren I. Spinal morphine administration reduces the fatty acid contents in spinal cord and brain by increasing oxidative stress. <u>Neurochm Res</u> 2007 Jan;32(1):19-25

Pacher P, Szabo C. Rose of poly (ADP-ribose) Polymerase 1 (PARP-1) in Cardiovascular Diseases: The Therapeutic Potential of PARP Inhibitors. <u>Cardiovasc Drug Rev</u> 25(3) 2353-60 Fall 2007

Pakhomov AG, Akyel Y, Pakhomova ON, Stuck BE, Murphy MR. Current state and implications of research on biological effects of millimeter waves: a review of the literature. <u>Bioelectromagnetics</u> 1998;19(7):393-413

Pakhomov A, Bojarinova J, Cherbunin R, Chetverikova R, Grigoryev PS, Kavokin K, Kobylkov D, Lubkovskaja R, Chernetsov N. Very weak oscillating magnetic field disrupts the magnetic compass of songbird migrants. <u>J R Soc Interface</u> 2017 Aug;14(133)

Pall ML. Electromagnetic fields act via activation of voltage-gated calcium channels to produce beneficial or adverse effects. <u>J Cell Mol Med</u> 2013 Aug;17(8):958-65

Pall ML. Scientific evidence contradicts findings and assumptions of Canadian Safety Panel 6: microwaves act through voltage-gated calcium channel activation to induce biological impacts at non-thermal levels, supporting a paradigm shift for microwave/lower frequency electromagnetic field action. <u>Rev Environ Health</u> 2015;30(2):99-116

Pall, ML. Microwave frequency electromagnetic fields (EMFs) produce widespread neuropsychiatric effects including depression. <u>J Chem Neuroanat</u> 2016 Sep;75(Pt B):43-51

Pall ML. (2018a). Wi-Fi is an important threat to human health. <u>Environ Res</u> 2018 Jul;164:405-416

Pall Martin L. (2018b). 5G: Great risk for EU, US and International Health! Compelling Evidence for Eight Distinct Types of Great Harm Caused by Electromagnetic Field (EMF) Exposures and the Mechanism that Causes Them. May 17, 2018. https://einarflydal.files.wordpress.com/2018/04/pall-to-eu-on-5g-harm-march-2018.pdf

Pall, Martin L. Microwave Electromagnetic Fields Act by Activating Voltage-Gated Calcium Channels: Why the Current International Safety Standards Do Not Predict Biological Hazard. https://olis.leg.state.or.us/liz/2014R1/Downloads/CommitteeMeetingDocument/35551

Panagopoulos DJ, Chavdoula ED, Nezis IP, Margaritiss IH. Cell death induced by GSM 900-MHz and DCS 1800-MHz mobile telephony radiation. Mutat Res. 2007 626 pp 69-78

Panagopoulos DJ. Chromosome damage in human cells induced by UMTS mobile telephony radiation. Gen Physiol Biophys 2019 Sep;38(5):445-454

Peleg M. Radio frequency radiation-related cancer: assessing causation in the occupational/military setting. Environ Res 2018 May;163:123-133

Phillips JL, Singh NP, Lai H. Eectromagnetic Fields and DNA Damage. Pathophysiology Aug 2009: 16 (2-3) 79-88

Physicians for Safe Technology. The First Report of 5G Injury   from   Switzerland. https://mdsafetech.org/2019/07/20/the-first-report-of-5g-injury-from-switzerland/ July 20, 2019

Physicians   for   Safe   Technology.   5G   "Mobile" Communications. https://mdsafetech.org/problems/5g/ Nov 4, 2019

Physicians for Safe Technology. 5G Telecommunications Science. https://mdsafetech.org/5g-telecommunications-science/ updated 1/11/20

Physicians for Safe Technology.   Senator Blumenthal Blasts FCC and FDA for No Research on 5G Safety. https://mdsafetech.org/2019/02/13/no-research-on-5g-safety-senator-blumenthal-question-answered/   Feb 13, 2019

Physicians for Safe Technology. Telecommunications Act of 1996. https://mdsafetech.org/telecommunications-act-of-1996/

Potekhina, I.L., Akoyev, G.N., Yenin, L.D., Oleyner, V.D., 1992. Effects of low-intensity electromagnetic radiation in the millimeter range on the cardio-vascular system of the white rat. Fiziol. Zh. [Former. Fiziol. Zh. SSSR] 78, 35–41. (in   Russian).   (https:// www.ncbi.nlm.nih.gov/pubmed/1330714).

Prasad M, Kathuria P, Nair P, Kumar A, Prasad K. Mobile phone use and risk of brain tumours: a systematic review of association between study quality, source of funding, and research outcomes. <u>Neurol Sci</u> 2017 May; 38(5):797-810

Prost, M., Olchowik, G., Hautz, W., Gaweda, R., 1994. Experimental studies on the influence of millimeter radiation on light transmission through the lens. Klin Oczna 96 (8–9), 257–259. ⟨https://www.ncbi.nlm.nih.gov/pubmed/7897988⟩.

Pryor WA, Squadrito GL. The chemistry of peroxynitrite: a product from the reaction of nitric oxide with superoxide. <u>Am J Physiol</u> 1995 May;268(5 Pt 1):L699-722

Ramundo-Orlando A.   Effects of Millimeter Wave Radiation on Cell Membrane – A Brief Review. <u>J of Infrared, Millimeter, and Terahertz Waves</u> Vol 31 pp 1400-11 (2010)

Redmayne M. International policy and advisory response regarding children's exposure to radio frequency electromagnetic fields (RF-EMF). <u>Electromagn Biol Med</u> 2016;35(2):176-85

Riva, C.E., Logean, E., Falsini, B., 2005. Visually evoked hemodynamical response and assessment of neurovascular coupling in the optic nerve and retina. Riva CE Prog. <u>Retin</u>

Eye Res 24 (2), 183–215. http://dx.doi.org/10.1016/j.preteyeres.2004.07.002. ⟨https://www.ncbi.nlm.nih.gov/pubmed/15610973⟩.

Russell, Cindy Lee. (2018a) 5 G wireless telecommunications expansion: Public health and environmental implications. Environ Res 2018 Aug;165:484-495

Russell, Cindy Lee. (2018b) Wireless Silent Spring. October 2018 Issue of the Santa Clara County Medical Association Bulletin. https://zero5g.com/wp-content/uploads/2018/10/SilentSpringAticle_color_pr2.pdf

Ryzhov, A.I., Logvinov, S.V., 1991. Early ultrastructural reactions in various parts of the visual analyzer in guinea pigs after thermogenic microwave irradiation. Arkh Anat. Gistol. Embriol 100 (7–8), 30–36. ⟨https://www.ncbi.nlm.nih.gov/pubmed/ 1843431⟩.

Sagar, S, Adem SM, Struhcen B, Loughran SP,, Brunjes ME, Arangua L, Dalvie MA, Croft RJ, Jerrett M, Moskowitz, Kuo T, Roosli M. Comparison of radiofrequency electromagnetic field exposure levels in different everyday microenvironments in an international context. Env International Vol 114, May 2018 pp 297-396

Sage, C., Carpenter D.O., 2012. BioInitiative Working Group, BioInitiative Report: A Rationale for a Biologically-based Public Exposure Standard for Electromagnetic Radiation at ⟨www.bioinitiative.org⟩ (31 December 2012).

Sage C, Burgio E. Electromagnetic Fields, Pulsed Radiofrequency Radiation, and Epigenetics: How Wireless Technologies May Affect Childhood Development. Child Dev 2018 Jan;89(1):129-136

Said-Salmon IH, Jeaii FA, Moustafa ME. Evaluation of Wi-Fi Radiation Effects on Antibiotic Susceptibility, Metabolic Activity and Biofilm Formation by Escherichia Coli 0157H7, *Staphylococcus Aureus* and *Staphylococcus Epidermis*. J Biomed Phys Eng 2019 Oct 1;9(5):579-586

Sainudeen SS. Electromagnetic Radiation (EMR) Clashes with Honey Bees. International Journal of Environmental Sciences Vol 1, No 5. 2011

Salford LG, Brun AE, Eberhardt JL, Malmgren L, Persson BRR. Nerve cell damage in mammalian brain after exposure to microwaves from GSM mobile phones. Environ Health Perspect 2003 Jun; 111(7); 881-883

Santini R, Santini P, Danze JM, Le Ruz P, Seigne M. Investigation on the health of people living near mobile telephone relay stations: Incidence according to distane and sex. (article in French). Pathol Biol (Paris) 2002 Jul;50(6):369-73

Santini SJ, Cordone V, Falone S, Mijit M, Tatone C, Amicarelli F, Di Emidio G. Role of Mitochondria in the Oxidative Stress Induced by Electromagnetic Fields: Focus on Reproductive Systems. Oxid Med Cell Longev 2018 Nov 8;2918:5076271

Sasaki, K., Sakai, T., Nagaoka, T., 2014. Dosimetry using a localized exposure system in the millimeter-wave band for in vivo studies on ocular effects. IEEE Trans. Microw. Theory   Tech.   62   (7). http://dx.doi.org/10.1109/TMTT.2014.2323011. ⟨http:// ieeexplore.ieee.org/document/6818422/⟩.

Sato, Y, Kojimahara N, Yamaguchi N. Analysis of Mobile Phone Use Among Young Patients With Brain Tumors in Japan. Bioelectromagnetics 38 (5) 349-355

SCHEER   Scientific   Committee   on   Health, Environmental and Emerging Risks.   Statement on emerging health and environmental issues (2018) https://ec.europa.eu/health/sites/health/files/scientific_committees/scheer/docs/scheer_s_002.pdf

Schuz J, Jacobsen R, Olsen JH, Boice JD Jr, McLaughlin JK, Johansen C. Cellular telephone use and cancer risk: update of a nationwide Danish cohort. J Natl Cancr Inst 2006 Dec 6;98(23):1707-13

Schuz J. Exposure to extremely low-frequency magnetic fields and the risk of childhood cancer: Update of the epidemiological evidence. <u>Progress in Biophysics and Molecular Biology</u> Dec 2011; Vol 107 Iss 3:393-342

Schwarze S, Schneider NL, Reichi T, Dreyer D, Lefeldt N, Engels S, Banker N, Hore PJ, Mouritsen H.   Weak Broadband Electromagnetic Fields are More Disruptive to Magnetic Compass Orientaton in a Night-Migratory Songbird (Erithacus rubecula) than Strong Narrow-Bank Fields. <u>Front Behav Neurosci</u> 2016 Mar 22;10:55

Shahin S, Singh SP, Chaturvedi CM. Mobile phone (1800 MHz) radiation impairs female reproduction in mice, *Mus musculus*, through stress induced inhibition of ovarian and uterine activity. <u>Reproductive Toxicology</u> Oct 2017, Vol 73, pp 41-60

Sharma, AB, Lamba OS, Sharma L, Sharma A. Effect of Mobile Tower Radiation on Microbial Diversity in Soil and Antibiotic Resistance. <u>International Conference on Power Energy, Environment and Intelligent Control (PEEIC)</u> Publisher: IEEE 2018

Sharma VP, Kumar NR. Changes in honeybee behavour and biology under the influence of cellphone radiations. <u>Current Science</u> Vol 98 No 10, 25 May 2010

Shcheglov VS, Alipov ED, Belyaev IY.   Cell-to-cell communication in response of E. coli cells at different phases of growth to low-intensity microwaves. <u>Biochim Biophys Acta</u> 2002 Aug 15;1572(1):101-6

Shinjyo T, Shinjyo A. Significant Decrease of Clinical Symptoms after Mobile Phone Base Station Removal – An Intervention Study 2014 http://nebula.wsimg.com/d1e65ba8eb587c44cba6164dfe f44ed2?AccessKeyId=045114F8E0676B9465FB&disposi tion=0&alloworigin=1

Singh Rilvani S, Saravanamuttu S. Impacts of radio-frequency electromagnetic field (RD-EMF) from cell phone towers and wireless devices on biosystem and ecosystem—A review. Biology and Medicine 2013 4(4):202-216

Simko M, Mattson MO. 5G Wireless Communication and Health Effects—A Pragmatic Review Based on Available Studies Regarding 6 to 100 GHz. Int J Environ Res Public Health 2019 Sep; 16(18): 3406

Singh R, Nath R, Mathur AK, Sharma RS. Effect of radiofrequency radiation on reproductive health. Indian Journal of Medical Research 2018 Vol 148 Iss 7 pp 92-99

Singh, S., Kapoor, N., 2014. Health implications of electromagnetic fields, mechanisms of action, and research needs. Adv. Biol. 2014 (2014), 24. http://dx.doi.org/10.1155/ 2014/198609. (https://www.hindawi.com/archive/2014/198609/).

Sivani, S., Sudarsanam, D., 2013. Impacts of radio-frequency electromagnetic field (RFEMF) from cell phone towers and wireless devices on biosystem and ecosystem—a review. Biol. Med. 4 (4), 202–2016.

Spector, A., 1995. Oxidative stress-induced cataract: mechanism of action. FASEB J. 9, 1173–1182. ⟨http://www.fasebj.org/content/9/12/1173.short⟩.

Soghomonyan D, Trchounian K, Trchounian A. Millimeter waves or extremely high frequency electromagnetic fields in the environment: what are their effects on bacteria? Appl Microbiol Biotechnol 2016 Jun;100(11):4761-71

Stewart DA, Gowrishankar TR, Weaver JC. Skin Heating and Injury by Prolonged Millimeter-Wave exposure: Theory Based on a Skin Model Coupled to a Whole Body Model and Local Biochemical Release from Cells at Supraphysiologic Temperatures. IEEE Transactions on Plasma Science 2006 Vol 34 No 4 Aug

Strestha M, Raitanen J, Salminen T, Lahkola A, Auvinen A. Pituitary tumor risk in relation to mobile phone use: A case-control study. Acta Oncol 2015;54(9):1159-65

Svendsen AL, Weihkopf T, Kaatsch P, Schuz J. Exposure to magnetic fields and survival after diagnosis of childhood leukemia: a German cohort study. Cancer Epidemiol Biomarkers Prev 2007 Jun;16(6):1167-71

Szmigielski S. Reaction of the immune system to low-level RF/MW exposures. Sci Total Environ 2013 Jun 1;454-455:383-400

Thannickal VJ, Fanburg BL. Reactive Oxygen Species in Cell Signaling. Am J Physiol Lung Cell Moi Physiol Dec 2000; 279(6): L1005-28

Thielens A, Bell D, Mortimore DB, Greco MK, Martens L, Joseph W. Exposure of Insects to Radio-Frequency Electromagnetic Fields from 2 to 120 GHz. Scientific Reports 8 Art nbr 3924 (2018)

Tillman T, Ernst H, Streckert J, Zhou Y, Taugner F, Hansen V, Dasenbrock C. Indication of Cocarcinogenic Potential of Chronic UMTS—modulated Radiofrequency Exposure in an Ethylnitrosourea Mouse Model. Int J Radiat Biol Jul 2010 86 (7) 529-41

Tkalec M, Malaric K, Pevalek-Kozlina B. Exposure to Radiofrequency Radiation Induces Oxidative Stress in Duckweed Lemna Minor L. Sci Total Environ 2007 Dec 15; 388 (1):78-89

Torgomyan H, Trchounian A. Bactericidal effects of low-intensity extremely high frequency electromagnetic field: an overview with phenomenon, mechanisms, targets and consequences. Crit Rev Microbiol 2013 Feb;39(1):102-11

UN #FridayFact: One in three spoonfuls of food depends on bees! https://www.unenvironment.org/news-and-stories/story/fridayfact-one-three-spoonfuls-food-depends-bees May 18 2018

Usselman RJ, Hill I, Singel DJ, Martino CF. Spin biochemistry modulates reactive oxygen species (ROS) production by radio frequency magnetic fields. PLoS One 2014 Mar 28;9(3):e93065

Usselman RJ, Chavarriaga C, Castello pR, Procopio M, Ritz T, Dratz EA, Singel DJ, Martino CF. The Quantum Biology of Reactive Oxygen Secies Partitioning Impacts Cellular Bioenergetics. Sci Rep 2016 Dec 20;6:38543

Valko M, Rhodes CJ, Moncol J, Izakovic M, Mazur M. Free Radicals, Metals and Antioxidants in Oxidative Stress-Induced Cancer. Chem Biol Interact Mar 10 2006; 160(1): 1-40

Van Ummersen, C.A., Cogan, F.C., 1976. Effects of microwave radiation on the lens epithelium in the rabbit eye. Arch. Ophthalmol. 94 (5), 828–834. http://dx.doi.org/ 10.1001/archopht.1976.03910030410012. (http://jamanetwork.com/journals/ jamaophthalmology/article-abstract/631798?appId=scweb).

Vernon ST, Coffey S, Bhindi R, Soo Hoo SY, Nelson GI, Ward MR, Hansen PS, Asress KN, Chow CK, Celermajer

DS, O'Sullivan JF, Figtree GA. Increasing proportion of ST elevation myocardial infarction patients with coronary atherosclerosis poorly explained by standard modifiable risk factors. Eur J Prev Cardiol 2017 Nov;24(17):1824-1830

Vignal, R., Crouzier, D., Dabouis, V., Debouzy, J.C., 2009. Effects of mobile phones and radar radiofrequencies on the eye. Pathol. Biol. 57 (6), 503–508. http://dx.doi.org/ 10.1016/j.patbio.2008.09.003. ⟨https://www.ncbi.nlm.nih.gov/pubmed/ 19036534⟩.

Waldmann-Selsam, Cornelia. Radiofrequency radiation injures trees around mobile phone base stations. Sci of the Total Environment Aug 2016 572:554-569

Wang Y, Guo X. Meta-analysis of association between mobile phone use and glioma risk. J Cancer Res Ther 2016 Dec;12(Supplement): C298-C300

Warille AA, Onger ME Turkmen AP, Deniz OG, Altun G, Yurt KK, Altunkaynak BZ, Kaplan S. Controversies on electromagnetic field exposure and the nervous systems of children. Histol Histopathol 2016 May;31(5): 461-8

WHO. World Health Organization. International Agency for Research on Cancer. Press Release: IARC Classifies Radiofrequency Electromagnetic Fields as Possibly Carcinogenic to Humans. May 31, 2011 https://www.iarc.fr/wp-content/uploads/2018/07/pr208_E.pdf

WHO, 1981. Environmental Health Criteria: Radiofrequency and Microwaves. In: Biologic Effects and Health Hazards of Microwave Radiation: Proceedings on International Symposium 1973. Warsaw, Oct 15–18, 1973.Sponsored by the WHO, US Department of Health, Education and Welfare, and The Scientific Council to the Minister of health and Social Welfare, Poland. 〈http://apps.who.int/iris/bitstream/10665/39107/1/9241540761_eng.pdf〉.

Wells PG, et al. Reactive Species (ROS) Formation, Oxidative DNA Damage and Repair in Teratogenesis. Birth Defects Research Part A Clinical and Molecular Teratology May 2015 103(5):3359-359

Wheeler T. 2016. National Press Club Luncheon https://www.press.org/sites/default/files/20160620_wheeler.pdf

Wiart J. Analysis of RF exposure in the head tissues of children and addults. Physics in Med and Biology 2008 Aug 53(13):3681-95

Williams, Janice 4/21/17 Cancer Linked to Cellphone Use, Italian Court Rules in Landmark Case. https://www.newsweek.com/cell-phone-italy-cancer-mobile-phone-brain-tumor-587704

Wiltschko R et al. Magnetoreception in birds: The effect of radio-frequency fields. <u>Journal of the Royal Society Interface</u> 2015 12(103)

Wu T, (2015a) Rappaport TS, Collins CM. The Human Body and Millimeter-Wave Wireless Communication Systems: Interactions and Implications. <u>2015 IEEE International Conference on Communications (ICC)</u> Jun 2015

Wu, T, (2015b) Rappaport TS, Collins, CM. Safe for Generations to Come. <u>IEEE Microw Mag</u> 2015 Mar; 16(2):65-84

Wyde, Michael, 2016. NTP Toxicology and Carcinogenicity Studies of Cell Phone Radiofrequency Radiation-Slide Presentation. National Toxicology Program. National Institute of Environmental Health Sciences, BioEM 2016 Meeting, Ghent, Belgium. ⟨https://ntp.niehs.nih.gov/ntp/research/areas/cellphone/slides_bioem_wyde.pdf⟩.

Yakymenko I, Tsybulin O, Sidorik E, Henshel D, Kyrylenko O, Kyrylenko S. Oxidative mechanisms of biological activity of low-intensity radiofrequency radiation. <u>Electromagn Biol Med</u> 2016;35(2): 186-202

Ye, J., Yao, K., Lu, D., Wu, R., Jiang, H., 2001. Low power density microwave radiation induced early changes in rabbit lens epithelial cells. Chin. Med. J. 114 (12), 1290–

1294. 〈https://www.ncbi.nlm.nih.gov/pubmed/11793856〉.

Yinhui P, Hui G, Lin L, Xin A, Qinyou T. Effect of cell phone radiation on neutrophil of mice. <u>Int J Radiat Biol</u> 2019 Aug; 95(8):1178-1184

Yokus, B, Cakir DU, Akdag MZ, Sert C, Mete N. Oxidative DNA Damage in Rats Exposed to Extremely Low Frequency Electro Magnetic Fields. <u>Free Radic Res</u> Mar 2005 39 (3): 317-23

Yu, Y., Yao, K., Non-thermal cellular effects of low power microwave radiation on the lens and lens epithelial cells. <u>J. Int. Med. Res</u>. 2010 38 (3), 729–736. http://dx.doi.org/ 10.1177/147323001003800301. 〈https://www.ncbi.nlm.nih.gov/pubmed/ 20819410〉.

Zalyuboxskaya. Biological Effects of Millimeter Wavelengths. 1977 Declassified by the CIA 2012

Zhang, X., Huang, W.J., Chen, W.W., 2016. Microwaves and Alzheimer's disease. <u>Exp. Ther. Med</u>. 12 (4), 1969–1972. http://dx.doi.org/10.1097/00019052-20011200000008.

Ziskin, M.C., 2013. Millimeter waves: acoustic and electromagnetic. <u>Bioelectromagnetics</u> 34 (1), 3–14. http://dx.doi.org/10.1002/bem.21750.

⟨https://www.ncbi.nlm.nih.
gov/pmc/articles/PMC3522782/⟩.

Zothansiama, Zosangzuali M, Lalramdinpuii M, Jagetia
GC. Impact of radiofrequency radiation on DNA damage
and antioxidants in peripheral blood lymphocytes of
humans residing in the vicinity of mobile phone base
stations. <u>Electromagn Biol Med</u> 2017;36(3):295-305

About the Author

Claudia Drake has a Master's degree in Public Administration from Eastern Washington University and a Master of Arts in Philosophy and Religion from the California Institute of Integral Studies, San Francisco, CA. She resides in a small town in southwest Washington state.

She can be contacted at thedangersof5G@gmail.com

 And www.claudiadrake.org

www.ingramcontent.com/pod-product-compliance
Lightning Source LLC
Chambersburg PA
CBHW031044160726
47991CB00005B/2016